Worthington Hooker

Natural History

For the Use of Schools and Families

Worthington Hooker

Natural History
For the Use of Schools and Families

ISBN/EAN: 9783744749640

Printed in Europe, USA, Canada, Australia, Japan

Cover: Foto ©berggeist007 / pixelio.de

More available books at **www.hansebooks.com**

NATURAL HISTORY.

FOR THE USE OF SCHOOLS AND FAMILIES.

BY

WORTHINGTON HOOKER, M.D.,

PROFESSOR OF THE THEORY AND PRACTICE OF MEDICINE IN YALE COLLEGE,
AUTHOR OF "HUMAN PHYSIOLOGY," "CHILD'S BOOK OF NATURE,"
ETC., ETC.

Illustrated by nearly 300 Engravings.

NEW YORK ·:· CINCINNATI ·:· CHICAGO
AMERICAN BOOK COMPANY

Entered, according to Act of Congress, in the year one thousand eight hundred and sixty, by HARPER & BROTHERS, in the Clerk's Office of the District Court of the Southern District of New York.

Printed at
The Eclectic Press
Cincinnati, U. S. A.

PREFACE.

THERE are many good books on Zoology, or Natural History, as it is commonly termed; but none are properly adapted to instruction in schools. Some of them are too popular in their character, and some, on the other hand, are too scientific, or, rather, contain too many of the details of science; while in all there is too much matter, so that the pupil is confused with the multitude of things brought to view, and therefore obtains definite ideas of but few of them. I have aimed in this book to avoid these defects. My object has been to cull out from the immense mass of material which Zoology presents *that which every well-informed person ought to know*, excluding all which is of interest and value only to those who intend to be thorough zoologists.

It seems to have been forgotten by most writers of text-books on the natural sciences that a book for common study should be very different from a book for reference. Their books are therefore cumbered with much that is not of any use to the great body of pupils. The true plan for instruction in schools requires that, while the class-book should contain, clearly stated, only that which all ought to know, the teacher should have some works on the subject of a more extended character, to which he can refer whenever occasion calls for it.

If a spirit of inquiry be awakened in the class (as it surely will be if the text-book be of the right stamp and the teacher use it aright), questions will occasionally be asked which will call for information that must be gathered from larger works, or perchance *from the teacher's own observation.* This leads me to say that no text-book is rightly constructed that does not excite this spirit of inquiry and observation on the part of both teacher and pupil. The more it does so, the more fully is the true object of teaching attained; for the communication of knowledge is by no means of so much importance as the imparting to the mind the power and the disposition to acquire it of itself. Especially is this true of such a study as Zoology, which presents to the pupil abundant material for observation on every hand, in the garden and in the field, on the land, in the water, and in the air.

I will mention here some of the books which the teacher may use with profit for reference in teaching Natural History. Carpenter's Zoology, Carpenter's Animal Physiology, Agassiz and Gould's Principles of Zoology, Cuvier's Animal Kingdom, Redfield's Zoological Science, Nuttal's Ornithology, Kirby and Spence's Entomology, Harris on North American Insects, Jaeger's Life of North American Insects, Jones's Aquarian Naturalist, Buckland's Curiosities of Natural History, Broderip's Note-book of a Naturalist, Harvey's Sea-side Book, Rennie's Insect Architecture, Brocklesby's Views of the Microscopic World. Any of these will be of great advantage to the teacher, but I would especially recommend Carpenter's Zoology, which constitutes two volumes in Bohn's Scientific Library. Redfield's Chart answers a good purpose in presenting to a class a bird's-eye view of the animal kingdom.

In order that Natural History may be taught efficiently, it is necessary that the pupil should have some knowledge of Physiology. It will be well for him, therefore, to go through my "First Book in Physiology" before entering on the study of this book, and better still would it be if he has also gone through my "Child's Book of Nature," in the Second Part of which are presented such views of Physiology and Natural History together as can be readily comprehended by children of nine or ten years of age. Throughout the present work I have been particular to develop the intimate connection existing between Physiology and Zoology, knowing that a neglect of this point would abate essentially from both the interest and the usefulness of the study.

The study of Zoology has as yet been but little pursued, and I will present here some considerations which will show that it ought to have quite a prominence not only in academies, but also in our common schools.

First, this study has a practical bearing upon many of the most valuable and extensive occupations of man, agriculture, horticulture, etc. Many animals share with man the fruits of the earth, and therefore it is important for him to know how far and in what ways to prevent their undue increase. Then, again, some animals live on those which are destructive to the fruits raised by man, and so are really serviceable rather than injurious to him. How many mistakes have been made for want of proper observation of the habits of such animals! Many a bird, for example, has been killed because he picked up a few grains or ate a small quantity of fruit, when he really was of great service to the farmer or gardener, because he devoured daily a large number of worms, the grain or the fruit being a very small portion of his food. A war was year after year waged by every cotton-grower against an insect which was supposed to be very destructive to the plant. But after a while it was discovered that a great mistake had been made—that another smaller insect did the mischief, and that the one which had been destroyed in such great numbers was really the cotton-grower's friend, for it lived by preying upon this smaller insect. One example more shall suffice, although great numbers of a similar character might be cited. It is stated by Buffon that there was once great danger that the island of Bourbon would be entirely devastated by locusts, but it was saved from this catastrophe by the knowledge which the governor had of a fact in Natural History. He happened to know that a bird in India, called the Grakle, was of great service in destroying the eggs and grubs of these insects, and he therefore had a large number of pairs of this bird imported into the island. They multiplied rapidly, and in a few years the locusts were exterminated. But now the grakles, their natural food having given out, fell to digging up and eating the seeds sown in the ground. The people thereupon were aroused against them, and even obtained the enactment of a law for their extermination. But in a few years they saw their error, for the locusts largely increased again. A new supply of grakles was obtained, and their preservation was secured by very rigid enactments. So high were the grakles in favor as locust-killers, that physicians were directed to proclaim that their flesh was unwholesome, to prevent the people from eating graklepie, of which they were very fond. "But this *extraordinary* care," says Carpenter, "was injurious. The birds soon again cleared the island of locusts, and destroyed the grubs which injure the coffee-plantations. But when this supply failed them, they proceeded to attack the corn-fields and orchards, and even killed the young of pigeons and other domestic birds. In order to restore the balance, a sort of Malthusian law was enacted to prevent their numbers exceeding the quantity of their legitimate food; and when thus kept in check, they continued to do good without any admixture of evil."

Such facts as these indicate the wide benefits which the science of agriculture may derive from accurate observation of the habits and relations of animals. The more minds there are brought to engage in such observations, the more facts will be gathered into the common stock of information. And as the accuracy and extent of the observations depend on proper education in the observer, it is important that the observing powers be trained early; and we may say, therefore, that the whole subject of the relation of animal to vegetable life, so important to the farmer and the gardener, will never be thoroughly understood till the study of Nature be made prominent from the very beginning of education.

As animals furnish man, to a great extent, with food and clothing, and a large variety of articles for use and ornament, an increased observation would undoubtedly increase the amount of resources obtained from the animal kingdom. We may go farther than this, and say, that if we had been ready to take hints from the structures which we find in animals, and those which are built up by them, many improvements in the arts might have advanced much more rapidly than they have done. For example, in the construction of optical instruments, a difficulty which Sir Isaac Newton thought never could be remedied, chromatic aberration, might have been remedied long before it was if that perfect optical instrument made by the Creator, the Eye, had been properly examined in relation to this point. So, too, paper might long ago have been made from wood, if the habits of that first paper-maker, the wasp, had been observed.

Another reason for making this study prominent is, that its connection with other studies is such that it contributes greatly to their interest and resources. This is true, for example, of Geography. It adds vastly to the interest of this study to have the pupil know familiarly how the various tribes of animals are distributed over the earth, and what relation this distribution has to climate, situation, etc. The connection between Zoology and Geology is of the most intimate character, as the pupil will see in the course of his study of this book. Then, too, Chemistry and Natural Philosophy, especially the latter, have many of their best illustrations in the composition and structure of animals, so that Zoology, with its relations to Physiology properly developed, will offer no inconsiderable additions to the interest of the two departments of science above named.

But the grand practical benefit to be derived from the study of Natural History, or, indeed, any of the natural sciences, is the discipline which it gives the mental powers. It cultivates the perceptive and reasoning powers together, thus forming that habit of *intelligent observation* which makes its possessor, as a matter of course, a person of extensive general information, and is an essential element of success in almost any pursuit in which he may engage.

In the present prevalent mode of conducting education the observing powers of the mind are, we may say, systematically neglected. A premium, even, is paid for their neglect; for the study of language, the execution of the processes of mathematics, and the memorizing of Geography, Grammar, etc., are allowed to have such exclusive possession in most of our school-rooms, that any disposition on the part of a pupil to attend to Zoology, or any of the natural sciences, must be repressed, if he wishes to maintain his standing in school. And even if such studies are admitted at all, they commonly have a very subordinate place in the general arrangement, and an examination for the purpose of determining the standing of the pupil is not extended to such studies, because they are not deemed essential, but only extraordinary and ornamental.

This strange neglect of these studies is seen even in our colleges. When a young man, for instance, enters Yale College, he is not supposed to know any thing of the natural sciences, or at least no knowledge of them is required as a qualification for admission. And after his admission, he is drilled in mathematics and the languages alone for two long years. The natural sciences are wholly excluded till his junior year, when he begins to attend to Natural Philosophy, and in his senior year he is taught, necessarily in a very hurried manner, in Chemistry, Mineralogy, and Geology. Yale College by no means stands alone in this respect, for very nearly the same is true of most of the colleges in this country, showing how little importance is attached to the study of the natural sciences as a part of the system of education.

All this is *radically* wrong. The natural sciences ought to have a place on an equality with the other studies, and from the outset. The child, when he begins to attend school, is interested in any thing that calls forth suitably that joint employment of his perceptive and reasoning powers which we call observation; and, therefore, with his first learning to read, natural objects should be made the subjects of instruction. All teachers who have used my "Child's Book of Common Things," and who, in connection with its use, have brought natural objects into the school-room for "object lessons," as they are termed, know by experience that the plan recommended is a feasible one. This is teaching science; in a small way, it is true, but yet teaching it, and laying a good foundation for farther instruction, not merely in the facts learned, but in the habits of observation which are formed. There are numberless facts about air, water, light, plants, animals, etc., which the youngest pupils can understand, if they are presented in the right manner. And the busy inquiries which they make after the reasons of the phenomena, and their appreciation of them, if stated simply and without technical terms, show that such teaching is not profitless. Children are better philosophers than is commonly supposed.

Beginning thus, the natural sciences should be made

prominent *throughout the whole course of education*, not only because they contain largely what is of practical use in many of the avocations of life, and what needs to be known to give any one the character of a well-informed man, but also because they are quite as efficient in disciplining the powers of the mind as the study of the mathematics and the languages. It is clear that they are essential to a *symmetrical* mental development, for when they are neglected the observing powers are not duly educated. And besides, while it is the peculiar province of the study of mathematics to promote exactness of thought and reasoning, it fails to give that exaltation and wide range of mind which the investigation of the grand general principles of nature, the traces of the power and wisdom of the Creator, tends to produce. Then, again, the study of the natural sciences aids the pupil in acquiring a knowledge of language, for natural objects and processes furnish a large proportion of the words in daily use, and the mathematics derive so much of their real interest from their numerous applications to the facts which natural science brings to view, that the one class of studies is auxiliary to the pursuit of the other. On the whole, then, we may say that the three classes of studies indicated should, for the most part, go on together, and that the only question should be in regard to the proportion of time which ought to be devoted to each.

Many other considerations might be presented in favor of making Zoology and the other natural sciences prominent in education, but I will notice but two of them. One is the fact that they open never-ending resources for agreeable mental employment.

The phenomena of nature are ever before us, and their variety is without limit. One, therefore, who has pursued the study of nature throughout his course of education will never be at a loss for fresh material for observation. Especially is this true of that science to which the pupil is introduced in the present work.

The only other consideration which I shall present is the moral effect of the early study of natural science. Ever varying views of the traces of the wisdom, power, and goodness of the Deity can not fail to lodge in the young mind sentiments and opinions, which will be apt to forestall successfully the arguments of skepticism that may be presented in after years. No mere general views can do this, though they are often relied upon; but the actual and definite knowledge which study and observation give is required to effect it. This benefit can hardly be overestimated. The preoccupation of the mind by clear and abundant evidence is a preventive measure of vast importance. Better is it thus to shut out error, than to permit its admission and then attempt to cast it out.

The author has in the course of preparation books on some of the other natural sciences—Natural Philosophy, Chemistry, etc.—having the same general plan which has been adopted in this work. His object is to aid in the introduction of these studies into the common school, as well as the academy and college.

The books which I have already prepared have been used in some schools as reading-books at the same time that they are used for study, and with marked success. The plan adopted is this. The class read the lesson, the teacher remarking upon it so far as is thought proper; and then the recitation is heard at such a time as will allow a sufficient interval for the study of the lesson. The benefit of this plan consists in making reading a more intelligent and interesting exercise than it commonly is, for it is thus necessarily the distinct object to have the pupils *understand* what they read. In regard to this I would remark, that text-books on almost every branch should be so constructed, both as to arrangement and style, that they can be used in the way indicated. Let me not be understood to mean that I would discard "reading-books" altogether, but I would not have reading taught solely by them.

I have subjoined to this book a full index, and also a glossary upon a new plan. Technical terms I have made it a point to explain whenever they are first introduced; and therefore, in the Glossary, instead of giving the explanation of any term, I refer simply to the paragraph where the explanation may be found.

W. HOOKER.

NEW HAVEN, May, 1860.

NATURAL HISTORY.

CHAPTER I.

CLASSIFICATION OF ANIMALS.

1. The Animal Kingdom has four grand divisions, or sub-kingdoms: the Vertebrates, the Articulates, the Mollusks, and the Radiates.

2. The animals of the vertebrate sub-kingdom have a frame-work, or skeleton of bones, inside, covered up by some of the soft parts of the body. In Fig. 1 (p. 14) you have the skeleton of man. You see that somewhat round box of bones which contains the brain; the column of bones, 24 in number, extending from this through the trunk of the body; the *pelvis*, consisting of a wedge-shaped bone supporting this column and two broad, flaring bones, m and l, on each side; the breast-bone, with the ribs extending from it to the column of bones in the rear, and the collar-bone, g, stretching from it as a prop to the top of the shoulder joint; the arm-bone, i, with the two bones of the forearm, n and o, and the numerous small bones of the hand; the thigh-bones; the bones of the leg, v and u, and those of the foot of about the same number with those of the hand.

3. That part of the skeleton in which man is like a great variety of other animals is the central column of bones, and this is therefore taken as the characteristic of the division including man and these animals. In Fig. 2 you have one of the bones of this column, a being its front part, and b the sharp rear part, termed the spinous process. It is the row of these rear sharp parts

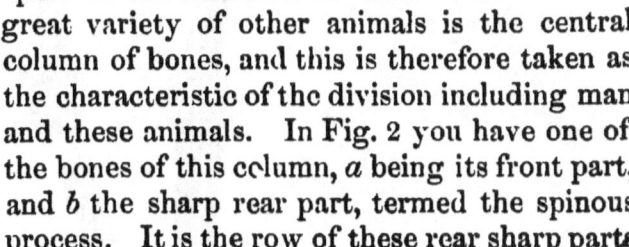

Fig. 2.—Single Vertebra.

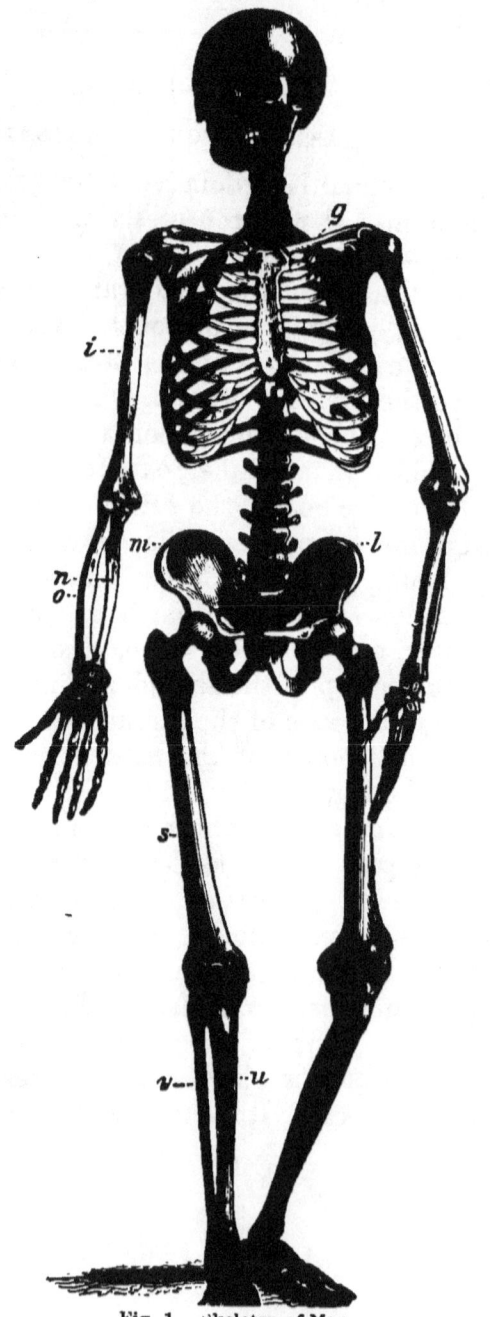

Fig. 1.—Skeleton of Man.

of the bones that you feel as you pass your finger up and down the middle of the back. Each of these bones is called a *vertebra* (plural *vertebræ*). Therefore all animals that have this column or chain of bones are called *vertebrate* animals. It is varied, as you will see, in different animals to suit different circumstances, and yet it is in essential points the same thing in all.

4. This vertebral column is found in all quadrupeds, as you see in this skeleton of a camel. The dark part

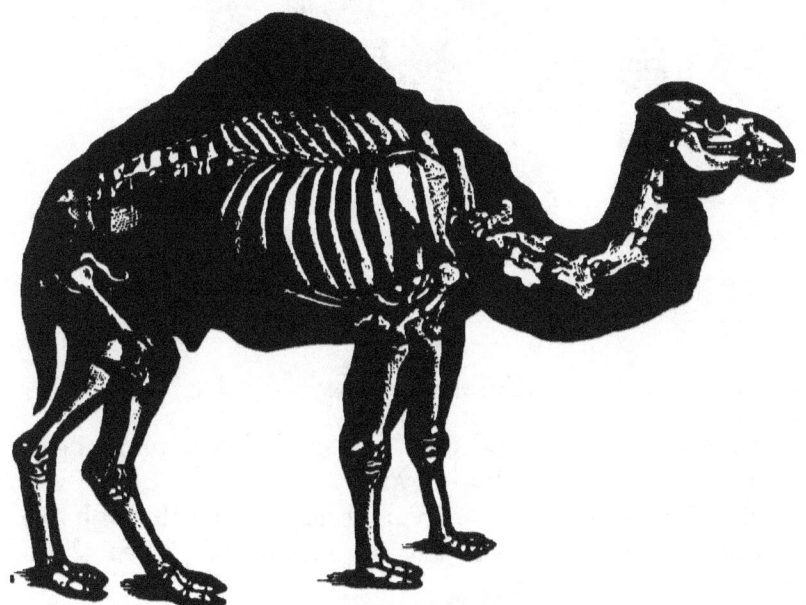

Fig. 3.—Skeleton of the Camel.

of the figure shows the full size of the animal. You observe that the spinous processes of the vertebræ of the back make a high, strong ridge. This is because to them are fastened the muscles that hold up the heavy neck and head.

5. Birds have this column, as you see in Fig. 4 (p. 16), the skeleton of an ostrich. Here there are very many more vertebræ in the neck than there are in the neck of

16 NATURAL HISTORY.

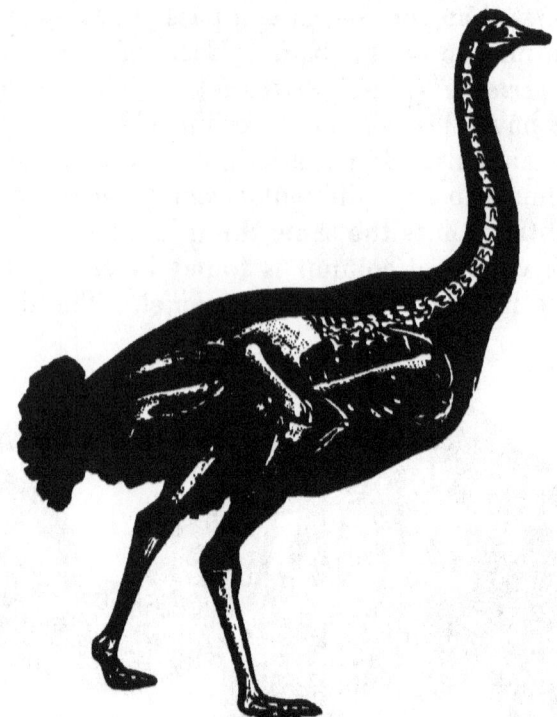

Fig. 4.—Skeleton of the Ostrich.

man, and near the head they are small, because the head is so small and therefore light.

6. In fishes the chain of vertebræ extends through the middle of the body, as you see in Fig. 5. Then there

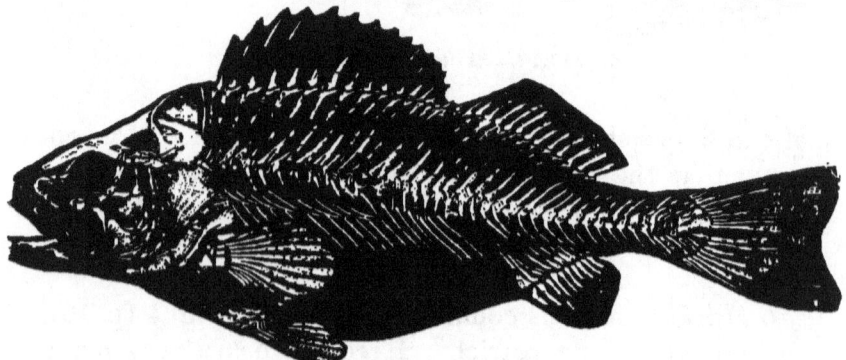

Fig. 5.—Skeleton of the Perch.

is another chain of bones of a slighter make along the back, their spinous processes being the frame-work of the fins of the back.

7. The turtle or tortoise tribe have the body covered with an upper and an under bony plate. But they have connected with the under side of the upper plate a true vertebral column. You see this in Fig. 6. The lower

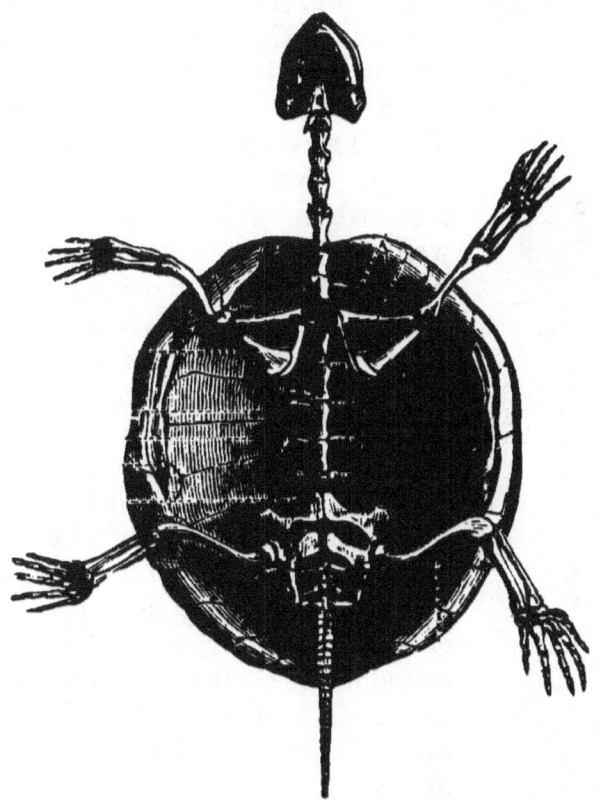

Fig. 6.—Skeleton of the Turtle.

plate is removed from this skeleton of a turtle to show the vertebral column in its whole length.

8. This chain or column extends out to the end of the turtle's tail. It is so with all the tails of four-footed animals. In the necks of birds, and generally of quadrupeds, and in the tails of the latter there is quite a free

motion among the vertebræ; while in the body of the animal the motion between them is slight.

9. In the snake tribe of Vertebrates the vertebræ are very numerous, and the motion between them is as free as in the tails of quadrupeds. Some species have over three hundred, while in man there are only twenty-four.

10. The skeletons of the different kinds of animals that I have mentioned differ from each other in many respects. For example, the fish has nothing in its skeleton that is like the bones of the extremities in man, and that of the serpent is composed merely of vertebræ, with very short ribs. There are some fishes that have no ribs. In the turtle, as you see in Fig. 6, the ribs spread out into broad plates, which, joined together, make its upper covering, termed the *carapace*.

11. While the differences are of extreme variety, the skeletons of all these animals agree in one thing—in having a vertebral column. They are, therefore, classed together as vertebrate animals.

12. Connected with this grand characteristic of this division of the animal kingdom there is another, viz., the arrangement of the great central organs of the nervous system. These are inclosed in the skull and vertebral column. The brain is in the skull, and the vertebræ contain the spinal marrow, which extends from the brain through the length of the body. Each vertebra has a round opening through it, as you see in Fig. 2. When, therefore, all the vertebræ are joined together, there is a tube-like passage through the column. In this lies the spinal marrow, or cord, as it is often called. In Fig. 7 you have

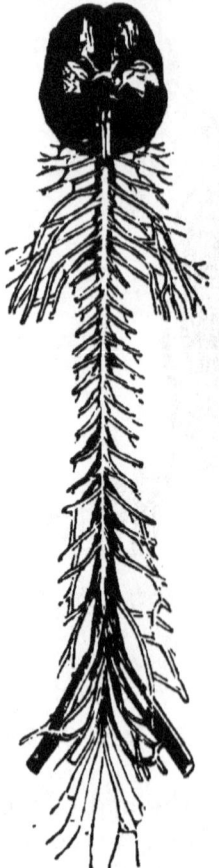

Fig. 7.—Brain and Spinal Cord of Man.

a representation of the brain and spinal marrow of man, with the beginnings of the nerves that branch out from them. Essentially the same arrangement exists in all the vertebrate animals.

13. The second grand division or sub-kingdom of animals is that of the *Articulates*. They have a jointed or articulated covering, as, for example, in the case of the lobster. They have no skeleton inside, as the Vertebrates have, but their coat of armor, as we may call it, is their skeleton. The muscles are all fastened to this. Thus, in the lobster, the muscles moving the claw have one end attached to some portion of the shell of the body, and the other to the shell of the claw.

14. The chief classes or tribes of the Articulates are the crab tribe, the worms, the spider and scorpion tribe, and the insects. In the crab tribe the jointed covering is very hard, being composed chiefly of a mineral substance—the carbonate of lime. In most of the insects it is very firm, and there is a marked resemblance in the claws of such insects as beetles to those of crabs and lobsters. Even in the worms the covering is firm compared with the soft interior parts.

15. The arrangement of the central organs of the nervous system of the Articulates is very different from that of the Vertebrates. There is no skull with a brain in it, and there is no spinal cord. There is a chain of little brains, as we may say, connected together by nerves, as represented in Fig. 8. Each of these is called a ganglion (plural ganglia). The first ganglion may be considered, for the most part, as corresponding with the brain in the Vertebrates, for the nerves from this go to the eyes and the other organs of sense.

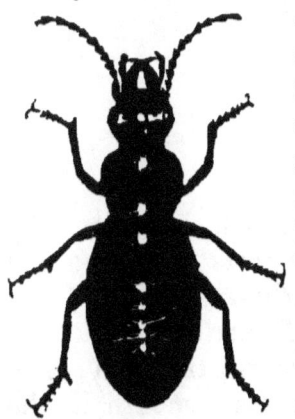

Fig. 8.—Nervous System of an Insect.

16. The third division of the animal kingdom is that of the *Mollusks*. This term comes from the Latin word *mollis*, soft. Mollusks are soft animals, most of them being inclosed in a hard shell, as the oyster, and all the varieties of shell-fish; and others being naked, as the slug. The central organs of the nervous system are ganglia variously arranged in the different orders of these animals.

17. The fourth sub-kingdom of animals is that of the *Radiates*. In Fig. 9 you have a representation of one

Fig. 9.—Star-fish.

of these animals, the star-fish, which will show you why they are called Radiates. You see parts extending like *rays* from the central portion. *Radius* is the Latin word for ray, and hence the name Radiate.

18. It is the upper side of the star-fish that you see in this figure. On the under side it has a mouth in the centre. The arrangement of its nervous system is sin-

gular. It is seen in Fig. 10. The place of the mouth is indicated at *a*. Around this is a nervous cord connecting together five ganglia, which are at the beginnings of the five arms of the animal. From each ganglion a nerve goes along each arm ending at its point in what is supposed by some to be a kind of eye. Though the animals of this sub-kingdom have great variety of form, the arrangement is essentially the same as that which you see in this animal.

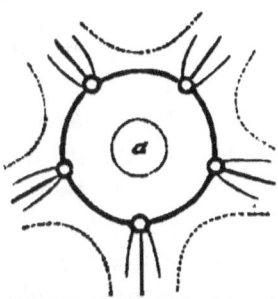

Fig. 10.—Nervous System of Star-fish.

19. These four sub-kingdoms are arranged in the order of their rank; the highest, or rather the most complicated, being placed first, and the simplest last. This is true of them in the general, and yet there are some in any one of the three lower divisions or groups that are higher in organization than some of the simplest in the one just above it. In the lowest group, the radiate, there are some animals which are nothing but a stomach with an apparatus to put food into it. The animals of one group are sometimes said to be more perfect than those of another; but this is not true, for the organization of every animal is perfectly adapted to its wants and its mode of existence.

20. There are many terms used in classifying the animals of each sub-kingdom, which you should understand at the outset. All animals that come from a common origin or parentage are said to belong to the same *species*. Thus all men descended from Adam, and therefore belong to one species, although they differ from each other in different quarters of the earth. These differences arise from accidental causes, as climate, food, habits, etc., and are not therefore *specific* differences. They make mere *varieties*, and not different species. So dogs and horses belong to two different species; but there are

varieties or breeds of dogs and horses, owing to accidental causes.

21. The distinction, then, between different species is a definite and fixed one. There can be no dispute about it in any case where the facts bearing on the question are all known; but it is not so with other distinctions, for they are not based upon specific and definite peculiarities, and may be varied by different classifiers. A *genus* includes many species that are alike in some things. Thus, the genus *canis* includes dogs, wolves, foxes, jackals, etc., which, though specifically different, are very much alike in their teeth, claws, and feet. Then a *family* includes genera (plural of genus); an *order* includes families; a *class*, orders; and, finally, orders are included in *sub-kingdoms* or departments. The terms division, tribe, and group are variously used by way of convenience. The term sub-class (under class) is sometimes used. It means a grand division of a class, as sub-kingdom means a grand division of a kingdom.

22. The Vertebrates have two grand divisions, the warm-blooded and the cold-blooded. The warm-blooded maintain a high temperature of the blood under varying states of the atmosphere. Thus, the blood of man is maintained at 98 degrees, even when the temperature of the surrounding air is 130 degrees below this. In the cold-blooded, on the other hand, the temperature of the blood is varied by that of the surrounding air or water. The fish when taken out of the water is of the temperature of the water, and therefore feels cold to our hands.

23. There are two classes of the warm-blooded Vertebrates: 1. Mammals, or Mammalia (from the Greek word μαμμα, *mamma*, a breast), animals that suckle their young; 2. Birds. The young of Mammals are born alive, and therefore Mammals are said to be *viviparous*, from the Latin words *vivus*, alive, and *pario*, to bear. Birds are called oviparous (*ovum*, egg, and *pario*), because their young are produced from eggs.

24. I divide the class Mammals into five sub-classes: 1. Bimana (Latin *bis* twice, and *manus* hand), two-handed animals. Man is the only representative of this sub-class. 2. Pedimana (*Pes* foot, and *manus*), foot-handed animals. This is the ape and monkey tribe. The name which I have given it is different from that which it commonly has in the classifications of zoologists, and the grounds of the change I will state when I come to speak particularly of this tribe. 3. Cheiroptera, hand-winged animals, or the bat tribe. This name is taken from two Greek words, χειρ, *cheir*, hand, and πτερον, *pteron*, wing. 4. Quadrupeds, or four-footed Mammals. Of these there are two divisions, the Unguiculata (Latin *unguis*, a nail or claw), and the Ungulata, from *ungula*, a hoof. 5. Cetacea, marine Mammals, or the whale tribe. These have neither hands nor feet. They were formerly classed with fishes, but although they are shaped like fishes, they have warm blood, and suckle their young, and have lungs and not gills. They, therefore, belong among Mammals, although they live in the water.

Questions.—What are the four grand divisions of the Animal Kingdom? Describe the skeleton of man. What is said of the central column of bones? Describe its arrangement in man. What is said of this column in quadrupeds? What of it in birds? What of it in fishes? What of it in the turtle tribe? What of it in the body, neck, and tails of various Vertebrates? What of it in tne snakes? What is said of the variety in the skeletons of different vertebrates? Why are they called Vertebrates? What is said of the nervous system of the Vertebrates? Describe the arrangement of the spinal marrow. What gives the Articulates their name? How are the muscles of the Articulates arranged? What are the chief classes of this sub-kingdom? What is said of the covering of these different tribes? Describe the arrangement of the nervous system of the Articulates. What is said of the Mollusks? Why do the Radiates have this name? What is the arrangement of their nervous system? What is said of the relative rank of the four sub-kingdoms? What of the use of the word perfect in regard to organization? Give the distinction between species and varieties. Give the various terms used in classification and their meaning. What are the grand divisions of the Vertebrates?

State the difference between them. Give the difference between the two classes of warm-blooded Vertebrates. What is the derivation of oviparous and viviparous? Name the sub-classes of the Mammals. What is said of the first class? What of the second? Of the third? Of the fourth? Of the fifth?

CHAPTER II.

MAN.

25. MAN is said to stand at the head of the animal kingdom. It is well that you should understand precisely what this means. We may consider every animal as a set of machinery, which is worked by means of the nervous system. In some animals this machinery is very simple, as in those which are nearly all stomach (§ 19). In others it is complicated. In man it is more so than in any other animal. For example, take that part of the machinery that is used in motion. Compare man with any animal in this respect. How many more motions he can make with his feet than a horse, or an ox, or a dog. The dog can walk, run, jump, and paw. To say nothing of other motions, observe in contrast the extreme varieties of motion of which the feet of man are capable in dancing.

26. There is no part of the machinery of the body in which man is so manifestly superior to other animals as in that of the hand. The variety of things that this machinery can do is so great, that you can get an adequate idea of it only by watching the motions of the hand in all the different kinds of work and play in which it engages.*

27. Look now at the instrument or machine itself. How simple it appears! You have merely a thumb and

---

* This and many other of the points in this chapter are quite fully treated in my "Child's Book of Nature," and "First Book in Physiology."

four fingers joined to the body of the hand; but observe how the thumb can be made to meet the tip of either finger, or to touch the tips of all of them at once, and how each finger can move independently of the others, or all can move together. Then observe, farther, in how many different ways the hand can take hold of different things, such as a pen, a whip, a rope, a string, an axe, etc.

28. What appears so simple when we look only at the outside, is found to be exceedingly complicated when examined within by the anatomist. The frame-work of this machine is made up of 32 bones, and there are numerous muscles with their cords or tendons. Then there are countless fibres branching from the nerves into these muscles. It is by these nerves that the mind in the brain works all this machinery.

29. Many animals have something like fingers, but none but man have any thing like thumbs except the monkey and ape tribe, and the opossum family; and in these the thumb is but a poor imitation of this organ in man.

30. While man is superior to all other animals in the *variety* of machinery in his body, there are some things in which some animals are superior to him. The horse, that is so inferior to man in the variety of his muscular movement, has better running machinery than he has. The monkey, the squirrel, the cat, etc., are better climbers. Fishes are better swimmers. And some animals have machinery which man does not possess at all, as flying machinery. The body of man, then, is superior to that of all other animals as a whole, but not in all respects.

31. The body of man is superior to that of other animals in some things besides those already mentioned. It is the only animal body that can maintain a perfectly erect position. The monkey can, indeed, stand and walk on its hind feet, or rather its foot-hands; but its position

is by no means perfectly erect, and it goes on all-fours except when compelled to do otherwise by its keeper.

32. There is superiority also in beauty of form and grace of movement. To make the comparison correctly, take the most beautiful and graceful of animals, and place them side by side with the most beautiful and graceful of the human race. Look now at form in detail. Take, for example, the upper extremity of man. Is there any thing in the limb of any animal to compare with it in its varied beauty of outline as it is placed in different positions? Observe, too, its graceful movements, and contrast their endless variety with the very limited grace of the corresponding limb of the inferior animal.

33. But in the face more than in any other part is seen this superiority both in form and movement. And when we look at the body as a whole, with its commanding erectness, the varied grace of all its parts as it moves, and its crowning head so full of the graces of expression, we realize that the human body is the only one that is a fit tenement of a soul made in the image of God.

34. This leads me to say that really the grand distinction between man and other animals is in the mind rather than in the body. He not only thinks more than any other animals do, but much of his thinking is wholly different from theirs. Even the most thinking of them know nothing about the difference between right and wrong, or about God; and you can not in any way teach them any thing in relation to such subjects.

35. As the mind of man is so superior to that of other animals, it can use more machinery than theirs can, and therefore more machinery is furnished it. For this reason man has a much larger brain than any other animal in proportion to the size of the body. The machinery of the hand is furnished to him because his mind requires it for the proper exercise of its powers on the world around. It would do no good to furnish a horse or a dog with a hand, for he would not know how to use it. Each ani-

mal is supplied with just the bodily machinery that its wants and capabilities require.

36. It is because the mind of man is not only superior to that of other animals, but is different in *kind* in some respects, that man has made and is continually making language. This no other animal has ever done. The inferior animals may have natural cries and signs, but they never agree to use artificial ones, and language is naught but a set of artificial signs. Some animals *imitate* spoken language, but they never make it.

37. For the same reason man is the only animal that makes tools, and some one proposed to designate man as a tool-making animal. I think that we may go so far as to say that other animals never use tools placed in their way except from imitation of man. And even the most knowing and imitative do but little at this. "An ape," says Wood, " will sit delighted by a flame which a chance traveler has left, and spread its hands over the genial blaze; but when the glowing ashes fade, it has not sufficient understanding to supply fresh fuel, but sits and moans over the expiring embers."

38. If we look at the mind of man alone we do not think of him as an animal. We think of him in this light only when we observe his bodily organization, and see its resemblance to that of the higher orders of animals, and even in some respects to that of the lower also. These two views of man are seen in the common expressions which are used. When we use such expressions as man and other animals, or man and the inferior animals, we have in view bodily organization. When, on the other hand, we use the expression man and animals, we have regard to those mental endowments which separate man entirely from animals. It is not in this view, but in the former, that the zoologist regards man in his classification.

39. Mankind are one species, as already stated in § 20. But there are certain varieties or races of men quite distinct from each other. The *Caucasian* race inhabits, for

the most part, Europe, the western part of Asia, and the United States. It is characterized by the oval shape of the face, a considerable variety of color both of the skin and the hair, and mental superiority. It is called Caucasian, from the Caucasian Mountains, in the neighborhood of which this race was at first settled. Even at the present day it is said that the external characteristics of this race are better developed in that locality than any where else, the Georgians and Circassians being the handsomest people in the world. The negro, or *Ethiopian* variety, I need not describe. The *Mongolian* race, of which the Chinese are the largest family, is characterized by prominent broad cheek-bones, a flat square face, small oblique eyes, straight black hair, a scanty beard, and olive skin. The *American* variety has high cheek-bones, large and bold features, except the eyes, which are sunken deeply in the sockets, hair generally black and stiff, and a copper complexion. In the *Malay* race, inhabiting the islands south of Asia, in the Indian and Pacific Oceans, the complexion is brown, the hair is black and thick, the forehead is low and round, the nose is full and broad with wide nostrils, and the mouth is large.

40. So great is the difference between these varieties, especially the Caucasian and the Ethiopian, that some believe that they came originally from different pairs. But the Bible declares that they were all descended from one pair, and almost all physiologists consider this to be also proved by a candid examination of facts. The different races of man are not more distinct from each other than the varieties of dogs and other animals. It is a remarkable fact that animals which remain wild are not apt to have varieties, while in those which are domesticated by man different breeds or varieties arise. Thus lions and tigers remain always the same, but dogs, horses, etc., have many varieties. So it is with man. Under the various influences to which he is subjected in society, in different ages and localities, varieties are produced.

41. The races of men may also be subdivided into varieties. Each nation has characteristics which are sometimes very marked. Thus the English and the Irish can ordinarily be readily distinguished at a glance. The Jews also have always been remarkably distinct from other nations. Then, too, we occasionally see an individual family with such striking peculiarities descending from father to son that we may call it a variety.

Questions.—What is said of the machinery in different animals? What of the variety of motion in the foot of man, and in his hand? What of the apparent simplicity of the hand as an instrument? What of its movements? What of its internal structure? What is said of the thumb? In what consists the chief superiority of the frame of man to that of other animals? In what respects are some animals superior to him? What is said of his erectness? What of his form and mode of movement? What of his face? What is the grand distinction between man and other animals? What is said of the machinery which the mind uses? What is said of language? What of making tools? What two views are taken of man, and to what modes of expression do these give rise? How many varieties are there of the human race, and what are they? Describe the Caucasian, the Ethiopian, the Mongolian, the American, the Malay. What is the testimony of the Bible as to their origin? Give the comparison between the varieties of the human race, and the varieties in animals. What is said of national and family varieties?

CHAPTER III.

FOOT-HANDED AND HAND-WINGED VERTEBRATES.

42. The sub-class which I call *Pedimana* is termed, in the common classifications of zoologists, the order *Quadrumana*, four-handed animals. It is the ape and monkey-tribe. I have already spoken in Chapter II. of the capabilities of the hand of man as an instrument. If we compare them with the very limited capabilities of the hand of the ape or monkey, we must agree with Sir Charles Bell, who says that " we ought to define the

hand as belonging *exclusively* to man." The chief object in the construction of the so-called hands of this tribe is to enable them to grasp the limbs of trees in climbing, in which they are greatly skilled. They are very imitative beings; but, even when they are subjected to long training, they can do but a few of the many things that can be done by the hands of man. On the whole, we may say that they have four members which partake in part of the character of a hand, and in part of that of a foot. It is for this reason that I have adopted the name of Pedimana, foot-handed. There is another reason for this in the fact stated by Dr. Carpenter, that one large division of this tribe have this resemblance to hands in only one pair of the extremities, and that the hinder pair. It is for this reason that he suggested the name which I have adopted, giving it less breadth of meaning, however, than I do. The suggestion is so good a one, that I wonder that he did not adopt it in his classification.*

* I may be considered by some as presumptuous in thus changing a name which has so long been retained in zoological classifications that it has almost acquired a right to its place by possession. But if the suggestion of Dr. Carpenter be a correct one, following it out fully can not only do no harm, but will certainly do good by placing the subject in its true light. If Sir Charles Bell is right in saying that no animal but man has truly a hand, and if the estimate which, in Chapter II., I have put upon this instrument, as fitly corresponding with man's mental capabilities, be correct, it is surely going very wide of the truth to call the hand-feet of the ape and monkey tribe real hands.

In this connection, I will remark on another change that I have made in the commonly received classification. Ordinarily, man is considered as one of the orders of the sub-class Unguiculata. But I have put him (§ 24) in a sub-class by himself, thus not only separating him more distinctly from other animals, as I think truth requires, but securing in other respects a more natural classification of the whole class of Mammalia.

In some classifications man is placed in even nearer relations to other animals than in the one ordinarily received. Thus, in that retained up to the present time in the British Museum, the first order of the class Mammalia is Primates, including man, apes, monkeys, bab-

43. There are three divisions of this sub-class ordinarily recognized: the Simiadæ, or monkey tribe of the Old World; the Cebidæ, or monkey tribe of the New World; and the Lemuridæ, which are found chiefly in the island of Madagascar, and to some extent in Africa and India. All these animals are inhabitants of tropical climates, and live chiefly on fruits, in getting which from trees most of them show greater agility than any other animals. They are disposed to gather in troops, a tree sometimes having nearly a hundred monkeys in its branches.

44. The Simiadæ are classed in three divisions: the *apes*, which have no tails; the *baboons*, that have very short ones; and the *monkeys*, that have long ones. I will notice some of the prominent species of each.

45. The Chimpanzee, Fig. 11, which is in shape more like

Fig. 11.—Chimpanzee.

oons, and bats, as the different *families* of the order, the second order being Feræ, or wild beasts. Such a classification is not merely incorrect, but ridiculous.

man than any other animal, is found in the west part of Africa. Its height is from four to five feet. It commonly goes on all-fours, but it walks occasionally on its hinder hand-feet, though not with the erectness of man. Its ears are very large, and it has long, black, coarse hair, which hangs in heavy whiskers about its cheeks. It climbs trees readily, sometimes for observation, and sometimes to gather food; and it makes a nest for itself by twining branches of trees together, in which it spends much of its time. Its strength is astonishing; it being able to break off branches which two men together can not bend.

46. The Orang-outang, Fig. 12, is an inhabitant of the

Fig. 12.—Orang-outang.

islands of Borneo and Sumatra. This is the largest of the apes, having been known to be in some cases over seven feet high. Its arms are of great length, reaching to the ground when it is erect. It can not stand as well

FOOT-HANDED AND HAND-WINGED VERTEBRATES. 33

as the Chimpanzee can, for it is so bow-legged that the soles of the feet turn in toward each other. Like the Chimpanzee, it is great at climbing, in doing which its long arms are very serviceable. When young it is very teachable, and has been taught to make its own bed, and to manage a cup and saucer and spoon tolerably well. Both the Chimpanzee and the Orang-outang have a gravity and apparent thoughtfulness which are quite laughable.

47. There are some smaller apes of an interesting character. The Agile Gibbon, so called from the agility with which it leaps from branch to branch, is a native of Sumatra. Its height is about three feet. A female of this species was some time since exhibited in London. She would leap over a distance of eighteen feet, and catch apples or nuts thrown up to her as she passed. As she leaped back and forth, which she did with great rapidity, she uttered a very loud but musical cry. She was a tame and gentle animal, and liked to be caressed.

48. I will notice but two of the many species of monkeys of the Old World. The Entellus, Fig. 13, is found in India. It preys upon serpents. In the attitude which you see here it steals quietly upon the serpent while it is

Fig. 13.—Entellus.

asleep, and seizing it by the neck, takes it to a stone, and knocks its head against it till it is dead. It then throws the snake to the young monkeys, who play with it as a kitten does with a mouse killed by the old cat. It is regarded with great reverence by the natives, and receives even divine honors from them. Splendid temples are dedicated to these monkeys; there are hospitals for their treatment when sick; fortunes are bequeathed for their support; and though the murder of a man is often punished only by a small fine, the killing of one of these monkeys is invariably punished with death. Thus cared for, they abound in great numbers, and though they enter houses to plunder eatables, their visits are regarded as a great honor.

Fig. 14.—Proboscis Monkey.

49. The Proboscis Monkey, Fig. 14, so called from the extraordinary projection of its nose, is a native of Borneo.

50. The baboons have very short tails. Their bodies are stout and thickset. The temper of most of them is very ferocious, and Cuvier says that he has seen several of the Mandrill species die of rage. Those species of baboons that live in Asia are of a much milder character than those found in Africa. There is only one locality in Europe where any of the Pedimana tribe are found, and that is the Rock of Gibraltar. One species of the baboon, improperly called the Barbary Ape, abounds there. It is probably not a native, but was originally introduced from the African side of the strait.

51. It is a remarkable fact that the baboons are the only Mammalia that exhibit bright colors upon their

skins. The Mandrill, Fig. 15, the largest and fiercest of the class, is prominent in this respect. Its colors are

Fig. 15.—Mandrill.

very brilliant and various. Being as tall as a man when erect, it presents a singular and formidable appearance. Its head is large, with very prominent eyebrows, and small, deeply-sunk eyes; the cheek bones are enormous, with large prominences on it of light blue, deep purple, and scarlet; its hair is an olive brown above and silvery gray below, but of a deep orange under the chin; the ears are violet-black, and the hinder parts of its body are a deep scarlet. This is Carpenter's description. The colors must vary in different cases, as I find them somewhat differently described by others.

X 52. The American monkeys are different species from those which we find in the Old World. Some of the particulars in which they differ from them I will mention. They are generally much smaller. The thumb is a very diminutive affair, and can not be brought in opposition to the fingers. In some cases it is wanting. The nostrils are wide apart, and open sidewise, while in the monkeys of Asia and Africa they are near together, and open downward. This makes a great difference in the aspect of the face. The monkeys of the Old World have cheek-pouches—that is, their cheeks are so loose and bag-like that they can stow away in them quite a quantity of nuts and other fruits as they gather them. These are not seen in American monkeys. The tails of American monkeys are in most species very long, and in many of them it is used as a sort of fifth hand in climbing. They are inhabitants of the northern half of South America. They are especially abundant in the vast forest-plains between the Orinoco and the Amazon. They live in trees, and pass from one tree to another with the same facility that squirrels do with us.*

53. I will notice but three of the many species. The Coaita Spider Monkey, Fig. 16, uses its tail, as you see,

Fig. 16.—Coaita Spider Monkey.

in climbing. It has been known to hang to a branch by it for some time after being killed by a shot. It uses its tail also to feel with, and to seize small things, such as eggs. For these purposes the end is destitute of hair, and is very sensitive. This animal is easily chilled, and

* Animals that live thus are said to be *arboreal* in their habits, from the Latin word arbor, tree

in cold weather it winds its tail around its body for warmth.

54. The Marmosets, of which you have one species in Fig. 17, are distinguished from other monkeys by their sharp and crooked nails. They are very skillful in capturing insects, which form a part of their food. Mr. Wood speaks of one in the Zoological Gardens in London which was very busy in catching flies. He caught some for it, and the little creature's eyes would sparkle with great eagerness as he saw Mr. Wood's hand moving toward a fly which had alighted out of its reach. In some of the species the tail is very elegant, from the different colors arranged in regular rings.

Fig. 17.—Marmoset.

55. The Howling Monkeys are larger than most American monkeys, and are morose in disposition. They have a sort of hollow drum connected with the windpipe, which gives great power to the voice in howling. They howl in concert at sunrise and sunset, often in the night, and also when a storm is threatened. The noise is described by travelers as astounding.

56. The Lemuridæ, or Lemurs (Latin, *Lemures*, ghosts), get their name from the fact that their movements are very noiseless, and are made mostly in the night. They live in troops, like the monkeys, clinging to branches of trees. Their food is various—fruits, eggs, insects, and birds. The posterior extremities, in contrast with monkeys and apes, are much longer than the anterior. The muzzle is pointed. The tail is commonly very long, but in some species is nearly wanting. The fur is usually fine

and silky. In the island of Madagascar, where these animals most abound, there are no monkeys. In Fig. 18

Fig. 18.—Ruffled Lemur.

you have the Ruffled Lemur of this island. In the Graceful Loris, Fig. 19, you have a Lemur that is found in In-

Fig. 19.—Graceful Loris.

dia and Ceylon. It is very skillful in capturing birds, which it does in the night, when they are asleep. Slowly and noiselessly advancing toward its victim, when it gets within reach of it, the Loris puts its hand toward it with a motion so slow as to be almost imperceptible, and then, with a motion quicker than sight can follow, it seizes its prey.

57. I will notice but one more species of the Lemurs. It is one whose skin is extended in a fold, like that of the Flying Squirrel, between its anterior and posterior

limbs. It is called the Flying Lemur. It has, however, like the Flying Squirrel, no power to fly upward; but this extension of skin merely enables it to take long sweeping leaps from one tree to another. It is a native of the Moluccas, Philippines, and other islands of the Indian Archipelago.

58. In the sub-class of *Cheiroptera*, or hand-winged Mammals (§ 24), we have the only animals of the class Mammalia that can really fly, that is, which can go upward in the air. The apparatus for flying is made up of a very delicate skin, without hair, on a frame-work of long slender bones. The bones are essentially the same that we find in the arm and hand of man, except that most of them are very much longer. This you can see by observing the skeleton of the bat in Fig. 20 in connection

Fig. 20.—Skeleton of the Bat.

with the skeleton of man in Fig. 1. Beginning at the shoulder, you see first the bone of the arm, then the forearm, and from the wrist extend the bones of the four fingers enormously lengthened. If the bones of the fingers of man were lengthened as much in proportion to his size, his fingers would be about four feet long. What answers to a thumb in the bat is a short projection with a hook upon it, as you see in the figure. Wood says of this arrangement that, " if the fingers of a man were to be drawn out like wire to about four feet in length, a thin membrane to extend from finger to finger, and another membrane to fall from the little finger to the ankles, he would make a very tolerable bat." He would need, however, vastly larger muscles than those which move his arm to work such extensive flying machinery.

59. The wing of a bat is a more extensive and perfect flying apparatus than that of any bird. Hence the exceeding rapidity of its movements. In his flight he is catching flies, musquitoes, and other insects. In his mode of getting a livelihood he is like the birds of the swallow tribe.

60. The eyes of the bat are small, and his vision is undoubtedly very poor. How, then, can he catch insects on the wing? It is because his other senses are very acute. He hears quickly. Especially is this the case with the Long-eared Bat, Fig. 21. The organ of smell, too, is quite extensive, particularly in some species. Then, too, the membrane of the wings is fully supplied with nerves, and is exquisitely sensitive. To prove this, Spallanzani put out the eyes of some bats, and then let them loose in his room, across which he had stretched strings in various directions.— The bats in no case flew against them, but

Fig. 21.—Long-eared Bat.

readily avoided these and other obstacles. Of course, they did this with the sense of touch alone, and that chiefly in their wings. They instantly knew in this way when they were coming near something besides air. The senses of smell and hearing would help them to determine whether this something was an insect or such a thing as a string.

61. The bats of temperate climates are, like the frogs and toads, in a torpid state through the winter, this being necessary simply because the insects upon which they live are gone. For this purpose they lodge themselves instinctively in some secret place where they will not be likely to be disturbed.

62. The species of bats are very numerous. Some of the species in tropical climates are quite large animals. The Vampire Bat of South America, Fig. 22, measures

Fig. 22.—Vampire Bat

two or three feet from tip to tip of the wings. It lives by sucking blood from different animals, which it does while they are asleep, and commonly without awaking them. The wound which it makes is very small, and yet it sucks from it quite a large quantity of blood.

63. The most singular species of bat is found in the

island of Java, called the Kalong Bat. Its wings expand to the extent of five feet. Its head is like that of a fox, as you see in Fig. 23. This animal belongs to that division of bats which live principally on fruits. They live, like monkeys, in troops on trees.—The division is a small one compared with the insect-eating bats.—Their wings are by no means as extensive in proportion to the size of the body, and they therefore fly more slowly, not needing the swift flight of the other division, as they catch no insects. As their eyes are large, they have not, probably, the sensitiveness in their wings which is so characteristic of the insect-eating bats.

Fig. 23.—Kalong Bat.

Questions.—What tribe are the Pedimana? What is the name usually given to them? State the reasons for the change of name. Give the substance of the note in regard to classification. What are the three divisions of the Pedimana? What are the three divisions of the Simiadæ? What is said of the Chimpanzee? Of the Orang-outang? Of the Agile Gibbon? Of the Entellus? Of the Proboscis Monkey? What is said of the baboons? Describe the Mandrill. State the differences between the American monkeys and those of the Old World. What is an *arboreal* animal? What is said of the Coaita Spider Monkey? Of the Marmosets? Of the Howling Monkeys? Describe the Lemurs and their habits. Where are they chiefly found? What is said of the Graceful Loris? Of the Flying Lemur? Describe the flying apparatus of the Cheiroptera. How does its frame-work compare with that of the hand and arm of man? What is said of the power of this apparatus? What are the habits of bats? What is said of their senses? Give the experiment of Spallanzani. How do bats pass the winter in temperate climates? What is said of the Vampire Bat? What of the Kalong Bat?

CHAPTER IV.

CARNIVOROUS QUADRUPEDS.

64. WE now come to Quadrupeds (*quatuor*, four; *pes*, foot), four-footed Mammals. This sub-class includes most of the animals of any size that walk on the ground. It has two great divisions — the *Unguiculata*, or clawed Quadrupeds; and the *Ungulata*, or hoofed Quadrupeds. In the Unguiculata there are five orders: 1. Carnivora (*caro*, flesh, *voro*, to devour). 2. Insectivora — Insect-eaters. 3. Rodentia (*rodo*, to gnaw). 4. Edentata (*e*, without, *dens*, tooth). 5. Marsupialia, so called on account of a *marsupium*, or pouch in the skin, in which the mother carries her young for some time after birth. The division Ungulata has two orders: 1. Pachydermata ($\pi\alpha\chi\upsilon\varsigma$, *pachus*, thick; $\delta\epsilon\rho\mu\alpha$, *derma*, skin), thick-skinned Quadrupeds, including elephants, horses, swine etc. 2. Ruminantia (*rumen*, a stomach or paunch), cud-chewing Quadrupeds, as oxen, deer, camels, sheep, etc.

65. The order Carnivora is divided into five families: 1. Felidæ (*felis*, cat), the cat tribe, including cats, tigers, lions, etc. 2. Canidæ (*canis*, a dog), including dogs, wolves, foxes, etc. 3. Mustelidæ (*mustela*, a weasel), weasels, otters, etc. 4. Ursidæ (*ursus*, a bear), the bear family, bears, raccoons, etc. 5. Phocidæ ($\phi\omega\kappa\eta$, *phokê*, a seal), seals, walruses, etc.

66. Many of the animals which we have already noticed have the power of living in whole or in part upon animal food, as, for example, man and some of the monkey tribe. But they can digest vegetable food also, and can even subsist wholly upon it. Even those which live on animal food alone, as some of the bats, eat insects and worms, and not the flesh of the larger animals, on which

the true Carnivora, with few exceptions, entirely subsist.

67. The animals of this order are readily distinguished from others by their teeth, which are formed for seizing, tearing, and cutting flesh, while those animals that eat grains and grass have their principal teeth formed for grinding. In Fig. 24 you have a representation of one side of the jaws of a carnivorous animal. The very long pointed teeth are called *canine* teeth, because they are so observable in the dog. The teeth in rear of these are mostly cutting teeth, the upper and lower going a little past each other so as to cut like scissors. Herbivorous (herb or vegetable eating) animals have grinding teeth in this rear part of the jaw.

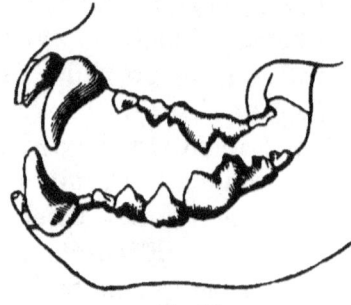

Fig. 24.

68. The digestive organs of this order are conformed to the nature of their food. As this is similar in quality to the substance of the animal itself, it does not require any complicated process to bring it into a fit state to nourish it. The stomach is therefore very simple and small, and the intestines are short; while in the grain and grass eating animals the digestive apparatus is complicated and extensive, it requiring, of course, much machinery to change into blood substances which are so unlike it as these articles of food are. I shall speak of this subject again when I come to the herbivorous Quadrupeds.

69. Some of the families of this order are not wholly carnivorous. And just so far as any admit vegetable food into their diet we see a corresponding variation from the true carnivorous character of the teeth and the digestive organs. The teeth, for example, lose to a greater or less extent their tearing and cutting character.

70. The animals of the first family, the Felidæ, or Cat tribe, are wholly carnivorous. They never eat vegetable food in their wild state, and eat but little of it when domesticated, as we know in the case of the common cat. The Felidæ, then, may be considered the *typical** family of this order. The animals which it includes are the most destructive of all the Mammalia, and the body is framed in every respect to conform to the carnivorous propensity. It has no unnecessary bulkiness, but is made as small as it can be, consistent with the required strength. Bone, and muscle, and sinew are well packed together, with but little fat. The limbs are short, for these animals need not to run so much as to leap in taking their prey. They have cushions or pads on their feet, so that they may approach their victims noiselessly. As they walk, their sharp claws lie back above these pads in their sheaths; but when they wish to use them, they thrust them forth from these sheaths by a very curious muscular apparatus. Their senses are acute, and they can see by night as well as by day. Their whiskers are very sensitive organs of touch, which are of service in passing through thickets or narrow places. The tongue is covered with almost horny points, directed backward. These, which every one has observed in the cat, are so large and strong in the lion and tiger, that a smart stroke of the tongue would strip off the skin from a man's hand. The chief use of these points is to enable the animal to scrape off all the flesh from a bone. The cat uses her tongue as

---

* This word, which is often used in works on Zoology, I will explain. In every natural group of animals there is always some one kind which exhibits the characteristics common to the group with more distinctness and perfection than any of the rest, and this is said, therefore, to be the *type* of the group. Thus, each genus has its typical species, each family its typical genus, each order its typical family, and each class its typical order. Then there is more or less variation from the *type*, and those which vary considerably from it are styled *aberrant* forms, from *erro*, to wander, and *ab*, from. So we speak of aberrant species, genera, etc.

a sort of curry-comb to clean her coat, and undoubtedly this member is put to the same use by other animals of this class in proportion to their cleanliness.

71. I will now proceed to notice some of the animals of this family. At the head of it, and of the wild beasts generally, stands the Lion. He is commonly called the

Fig. 25.—Lion, Lioness, and Cubs.

king of beasts, both for his noble and commanding air, and the power concentrated in his comparatively small frame. No animal, however large, dare attack him. He is found in Africa, and on the Continent of Asia, in India, Persia, and Arabia. He preys upon antelopes, heifers, zebras, gnoos, etc. There is such prodigious strength in the muscles of his neck and jaws that he can carry off a heifer as easily as a cat can a rat. He generally waits in ambush for his victim, or creeps like a cat insidiously and noiselessly toward it, and, when sufficiently near, at one bound secures it with his teeth and claws, uttering, at the same time, his terrific roar. He is not properly styled king of the *forest*, for he frequents burning desert plains, or places covered with low brushwood. He commonly sleeps in the day, and at night rouses to search for prey. A thunder-storm, so common in the night in Southern Africa, seems to excite him to unwonted activity, and, mingling his roar with that of the thunder, he rushes upon his terrified and confused prey without his usual stealthiness. The Lioness is smaller than the Lion, as you see in Fig. 25 (p. 46), and is destitute of the mane which gives him so dignified an appearance. The cubs, of which there are commonly from two to four, are as playful as kittens. Mr. Wood says that he had two cubs, larger than cats, placed in his arms, and found them "almost unpleasantly playful."

72. The Tiger, Fig. 26, is found only in Asia, chiefly in

Fig. 26.—Tiger.

Hindostan. It is a splendid animal, three feet high and eight long, having black stripes on a ground of reddish yellow. Tiger-hunts are among the favorite sports of India. The hunters go forth armed with rifles, in a sort of carriage or frame on the backs of elephants trained for the purpose. There is much danger in the sport, for the Tiger often springs up upon the elephant, and reaches the hunters.

73. The Leopard is a native of Africa, India, and some of the Indian islands. It is a very active and graceful animal. It is arboreal (§ 52) in its habits, and monkeys form a part of its prey. It has black spots in rosette shape, on a ground of pale yellow. The Ounce, a native of India, has sometimes been confounded with the Leopard; but it has less regular marks, a rougher coat, and a tail almost bushy.

74. The Jaguar of America, Figure 27, is much like the Leopard of the Old World, but it is larger. It is arboreal, and chases the monkeys which are so abundant in the forests of South America. The Puma, called usually in this country the Panther, is another animal of the same sort, found extensively diffused in both parts of the American continent. It is sometimes termed the American lion, from its uniformity of color, which is a silvery fawn.

Fig. 27. —Jaguar.

75. Of the Lynxes there are several species, some in Europe, some in Asia and Africa, and others in America. The Canada Lynx, Fig. 28 (p. 49), is remarkable for its gait, going by successive leaps with the back arched.

CARNIVOROUS QUADRUPEDS. 49

Fig. 28.—Canada Lynx.

Fig. 29.—Civet Cat.

There is a very large trade in the skins of this animal.

76. The Civet Cats, which are found in the northern part of Africa, chiefly in Abyssinia, are all remarkable for a pouch near the tail containing a perfume which is quite an article of commerce. A representation of one species you have in Fig. 29.

77. The Ichneumons are singular animals. Having long bodies and short limbs, they creep into very narrow places, and run their slender snouts into every crevice in search of their food, which consists of snakes, lizards, crocodiles' eggs, etc. The Egyptian Ichneumon, or Pharaoh's Rat, Fig. 30, is often domesticated

Fig. 30.—Egyptian Ichneumon.

in houses in Egypt, that it may destroy the snakes and other reptiles that so often infest them in that country.

78. The domestic Cat is so well known that I need to say little about it. The species has many varieties, though not as many as there are among the dogs. The Cat has a strong attachment to localities, but seldom man-

ifests that attachment to persons which is so strong a characteristic of most dogs. The domestic Cat was formerly thought to be the same with the Wild Cat, but they are proved to be distinct species. The difference in their tails may be seen in Fig. 31, that of the domestic Cat, 1, being long and tapering, and that of the Wild Cat, 2, short and bushy.

Fig. 31.—Cats' Tails.

Questions.—What are Quadrupeds? What are the two grand divisions of this sub-class? Give the names of the orders of the Unguiculata, and their derivations. Give those of the orders of the Ungulata, and their derivations. What are the families of the order Carnivora? What is said of the Carnivora in comparison with the animals already noticed? What is said of the teeth of the Carnivora and of the Herbivora? What is said of the difference in their digestive organs? What is said of some families of the Carnivora which are not wholly carnivorous? What is the typical family of this order? What is the meaning of *typical,* and of *aberrant?* What is said of the structure of the Felidæ? Of what use are the pads on their feet? What is said of their senses? What of their tongues? Where is the Lion found? Describe his appearance and habits. What is said of the Lioness? What is said of the Leopard? Of the Jaguar? Of the Puma? Of the Lynxes? Of the Civet Cats? Of the Ichneumons? Of the Cat?

CHAPTER V.

CARNIVOROUS QUADRUPEDS—*continued.*

79. THE second family of carnivorous Quadrupeds is the dog family, including dogs, wolves, foxes, etc. The dog species in this family exhibits more striking varieties than any other species of animal. There can hardly be a wider difference between two animals of the same family than we see between King Charles's Dog, Fig. 32, and the fierce Bloodhound, Fig. 33 (p. 51). Then we have the large and noble Newfoundland Dog, the stout Mas-

CARNIVOROUS QUADRUPEDS. 51

Fig. 32.—King Charles's Dog.

tiff, the slender and swift Greyhound, the pugnacious Bulldog, the brisk little Terrier, the Foxhound, Beagle, and Pointer used in hunting, etc. The differences, you observe, are as wide in disposition and habits as in form, size, and color. Now all these varieties, it is agreed by all zoologists,

Fig. 33.—Bloodhound.

came from one source, though exactly what was the character of the original undomesticated dog is not settled.

80. The cause of the wide range of varieties in this species is the influence of domestication referred to in § 40. The degree of domestication is greater in the dog than in any other case. No other animal is so thoroughly the *companion* of man. Cuvier says that the dog is the only animal that has followed man through every region of the earth. His attachment to his master is peculiar, and is seldom seen in other animals in the same degree. The contrast between the cat and the dog in

this respect is very marked, the cat being much attached to place, and little, if any, to persons.

81. The differences between some of the *varieties* of dogs are greater than those existing between different *species* of some animals. The Greyhound and the Bulldog, for example, are more unlike than the Lion and the Tiger, two species of the cat tribe, and vastly more than the Tiger and the Leopard. But the characteristics of these species remain fixed age after age, because the influence of domestication is not brought to bear upon them. Even the markings on the skins of such wild animals remain unchanged from generation to generation. Stripes and patches are therefore, in some of them, made the basis of distinguishing different species, while in the domesticated animals nothing is more common than changes of color.

82. The differences between the varieties of man are no greater than those between the varieties of the dog, the companion of man. And if domestication can produce these varieties in the one case, they surely can in the other, where it has a still greater influence. The doubts, then, existing in the minds of some in regard to the single origin of the human race are unfounded, and the account given in the Bible is proved true by an observation of facts.

83. Although the Wolf, Fig. 34 (p. 53), belongs to the dog family, dogs seem to be its natural enemies. While the smaller flee from it in terror, the stronger pursue and kill it. And yet it is thought by some that the original dog was a Wolf; and it is asserted that, though this animal is so fierce, it can be tamed when young, and is then as susceptible of attachment to man as the dog is. Wolves commonly hunt in packs or bands, and are very crafty in their modes of taking their prey. Like other wild beasts, they are exterminated as man cuts down the forests and builds his habitations. In the early settlement of this country they abounded even in the states on

Fig. 84.—Wolf.

the Atlantic coast, and they were not wholly exterminated till recently. The story of Putnam and the Wolf is familiar to every one. They were extirpated in England about 1350, in Scotland in 1600, and Ireland in 1700. They still abound in various parts of Europe and Northern Asia, and destroy great numbers of domesticated animals, as is shown by a report made in 1822 to the Russian government in regard to the district of Livonia, a tract of country about 250 miles long by 150 broad. The animals stated as having been destroyed by wolves are as follows: horses, 1841; cattle, 1807; calves, 733; sheep, 15,182; lambs, 726; goats, 2545; kids, 183; swine, 4190; young pigs, 312; dogs, 703; geese, 673; fowls, 1243. The Wolf is a gaunt but strong animal, with a skulking gait, and his aspect is marked by mingled ferocity, cunning, and cowardice. There are several species of wolves, especially in America, but their habits and character are very much the same.

84. The Fox, Fig. 35 (p. 54), is characterized chiefly by its pointed muzzle and its bushy tail. Its cunning is also proverbial. It is usually concealed in the daytime either in a burrow that it has made, or in one that it has

Fig. 35.—Fox.

found, and comes forth stealthily at night in search of its prey, which consists of fowls, rabbits, etc. It is a great robber of the hen-roost. Though a slender animal, the Fox is very muscular, and has great speed. This, with the cunning which it exercises in its various expedients for escape, renders the fox-chase very exciting, and it is one of the grand sports of English noblemen. Besides the common Fox, there are many other species. The Arctic Fox, which is found only in the extreme north, is remarkable for the changes which its hair exhibits. In summer it is of a dusky ash color, but in winter it turns white, and becomes fuller and thicker, even covering the soles of the feet.

85. The Jackal, Fig. 36 (p. 55), is found in North Africa, Persia, and India. It is somewhat like the Fox in appearance, though it has not so bushy a tail. It is like the wolf, however, in its habits. Jackals, like wolves, hunt in packs. They are concealed during the day, and come forth at night filling the air with their shrieks, which all describe as being horrid. They are very useful in the Eastern countries as scavengers, devouring the offal which the uncleanly inhabitants cast out of their

Fig. 36.—Jackal.

Fig. 37.—Striped Hyæna.

houses, and thus often save them from pestilential diseases.

86. The Hyænas, of which one species is represented in Figure 37, are found in Asia and Africa. They are generally classified in the dog family, though there is some question as to the place in which they belong.—They are exceedingly ferocious, and live chiefly upon animals which they find dead. They will even devour the human body, and are seen in large numbers in the neighborhood of armies, ready to eat the bodies of the slain. They are among beasts what the vultures are among birds, and, like the jackals, are very useful as scavengers. The rear parts of the Hyæna are small, and hence its shambling gait; but there is great strength in the fore part of its body and in its jaws. It can readily crush with its teeth the thigh bone of an ox.

87. The Weasel family (mustelidæ) includes the Weasels, Martens, Skunks, Otters, etc. These animals are, for the most part, quite small, but they are very sanguinary in their habits. They generally strike the neck of their victims just behind the ear, piercing the large blood-vessels, or drive their teeth into the skull. When they have

once seized their prey, which is a rabbit, or rat, or bird, or some reptile, they never let go their hold. Few animals equal them in agility and address. As they have such long, slender, flexible bodies, and creep stealthily toward their prey on their short legs, they have been sometimes called *vermiform*, worm-like, Carnivora. They are nocturnal in their habits, spending the day concealed in hollow trees, holes in walls, or in burrows, and gliding forth at night after their prey. Some of the most beautiful furs are obtained from this family, as the Sable and the Ermine. Most of these animals have a strong odor. Some of them are exceedingly offensive.

88. The common Weasel, Fig. 38, exemplifies the general shape of the whole tribe, of which it is the smallest. This animal is so effective in exterminating rats and mice, that the farmer can well afford to let him steal now and then an egg or a chicken, which it will never do so long as any rats or mice are to be found on the premises.

Fig. 38.—Weasel.

89. The fur of the Sable is very valuable. Great numbers of this animal are taken by hunters in Siberia, and are a considerable article of the Russian trade. The fur of the Pine Marten comes next in value. Many other furs are furnished by this family. The fur of the Ermine was formerly used in England to line the robes of judges and magistrates, and was, therefore, often referred to figuratively as emblematical of the purity which should belong to such persons.

90. The Skunk genus, of which there are several species, found only in America, belongs to this family. The common Skunk is about the size of a cat. The offensive fluid which it can throw upon any that attack it is contained in two sacs near the tail. Like the Woodchuck,

in the Northern States it retires to its burrow in the autumn to sleep through the winter.

91. The Otters form a somewhat aberrant genus of the Weasel family. They differ from the other genera in being aquatic, their prey being for the most part in the water. Their paws are fitted for swimming, which they do with great celerity. Their fur is close, short, and fine, so that it may not interfere with their progress in the water, and they are provided with a nictitating (winking) membrane which can be drawn over the eye for defense, it being transparent enough to allow the animal to see through it. There is considerable resemblance in these animals to the seals soon to be noticed. There is one species found on the northwest coast of America, and on the opposite or northeast coast of Asia, which has this resemblance strongly marked. Its tail is short, and its hind feet form very broad paddles, and are situated far back for convenience in swimming.

Questions.—What are included in the second family of the Carnivora? What is said of the varieties of the Dog? What is said of the influence of domestication? How are the Dog and Cat contrasted? How does the difference between the varieties of dogs compare with that between the species of some animals? What is said in this connection of the varieties of the human race? Describe the Wolf and its habits. What is said of its relation to the Dog? What of its extermination? Of its ravages? What are the characteristics and habits of the Fox? What is said of the fox-chase? For what is the Arctic Fox remarkable? What is said of the Jackal? What are included in the Weasel family? What is said of their structure and habits? What is said of the common Weasel? What is said of the furs that come from this family? What is said of the Skunk? What of the Otters?

CHAPTER VI.

CARNIVOROUS QUADRUPEDS—*concluded*.

92. THE family of Ursidæ, the Bear tribe, includes the Bears, Raccoons, Badgers, etc. These are said to be Plantigrade animals (*planta*, sole, and *gradior*, I walk), because, like man, they apply the sole of the foot to the ground in walking. The families of the Carnivora already noticed are, on the other hand, said to be Digitigrade (*digitus*, finger or toe, and *gradior*), because they walk on their toes; the bone which corresponds to the heel-bone in man really extending quite up the leg. You can see how this is if you compare the skeleton of the camel, which is a Digitigrade animal, with that of man, in Figs. 1 and 3. To make the comparison clear, begin at the hip or shoulder joint of the camel, and go down to the feet, observing the corresponding bones in man.

93. Although this family is placed among the Carnivora, most of the species live partly on vegetable food, and some live almost entirely upon it. They may be said to be nearly, if not quite, omnivorous (*omnis*, all, and *voro*, to eat). Most of them are expert in climbing. They conceal themselves in caves, holes, and hollow trees; and it is in such places that they spend the winter in a state of partial torpidity. The genus Ursus, or Bear, is the type of the family. There are eight species: three in Europe—one of which, the Polar Bear, is common also in America; one in the mountains of India; one in Java; one in Thibet, and three in North America. The body and limbs of the Bear are massive, and are covered with shaggy hair. Its five toes have strong claws, suited to digging. In very cold countries bearskins are of great use in making coverlets and articles of clothing. Leath-

er, also, is made from them for harnesses. The Brown Bear of Northern Europe yields so many benefits to the people of Lapland that they call it "the dog of God."

94. The Grizzly Bear of North America, Fig. 39, is

Fig. 39.—Grizzly Bear.

the most fierce and powerful of the Bears. Among the Indians it is regarded a great feat to kill one of them, and he who does this is permitted to wear a necklace of its claws as a decoration. Although very clumsy, it climbs trees readily, which it does to get at the honey in the nests of wild bees. It lives on roots, berries, and juicy plants, and, when it can do so, will devour a pig, a sheep, or a calf.

95. The Polar Bear, Fig. 40 (p. 60), is entirely white, except the claws and the tip of the nose, which are black. It lives chiefly upon seals, which it hunts both in the water and on the ice. With its stout claws, and its long hair about its feet, it runs rapidly over the smoothest ice, and even climbs up the sides of icebergs. Sometimes these bears float off to sea on fields of ice, and in this

Fig. 40.—Polar Bear.

way they have been known to emigrate from Greenland to Iceland, and there find luxurious living in the flocks and herds of the inhabitants, a change from their customary seal diet which was very grateful to them.

96. The other animals of this family which I shall notice are much smaller, and belong to genera more or less aberrant. The Raccoon, Figure 41, is about the size of a Fox. Like the Bear, it has sharp claws and climbs trees. It sleeps in its hole by day, and prowls at night for its food, which consists of small quadrupeds, birds, eggs, insects, roots, etc. It is very

Fig. 41.—Raccoon.

CARNIVOROUS QUADRUPEDS. 61

dexterous in opening oysters. It bites off the hinge, and scrapes out the oyster with its paw.

97. The Badger, Fig. 42, is found throughout Europe and Asia. It has often been made the subject of a cruel sport, teasing with dogs, and hence the common term "badgering." Its food is various. It is very fond of honey, and attacks the nests of wild bees, which it does with impunity; for its skin is so tough and its hair is so thick that the bees "might as well sting a barber's block." Its hair is extensively used in making brushes, and the skin is used for holsters and the coverings of traveling trunks. There is an American Badger somewhat like that of the Old World.

Fig. 42.—Badger.

98. The Wolverine, or Glutton, Fig. 43, is a native of the Arctic regions of both continents. It has been called the Quadruped Vulture, because it sometimes preys on the dead bodies of animals. It does great damage to the fur trade. When it finds the hunter's traps set for the martens, it takes the bait, which is a bit of venison or a partridge's head, or, if there be martens in the traps, it tears them in pieces, and buries them here and there in the snow. It is said that the Wolverines do not eat the martens, but the cunning foxes on the watch readily scent them out and devour them.

Fig. 43.—Wolverine.

99. The Kinkajou, Fig. 44, is found in South America.

Fig. 44.—Kinkajou.

It has been called the Honey Bear, because it is so fond of attacking the nests of the wild bee, licking out the honey from the cells with its long tongue. It is also very expert with its tongue in catching flies and other insects. Its tail it uses, like the Spider Monkey of the same country, in climbing. It is easily tamed, and is as playful as a cat.

100. The family Phocidæ (φωχη, *phokè*, a seal) are Quadrupeds, and yet they are fitted to live in water as well as on the land. There was an approach to this in the Otters, § 91. Seals and other animals having a similar mixture of terrestrial and aquatic habits, are often termed *amphibious* animals, from αμφι, *amphi*, both; βιος, *bios*, life.

101. The limbs of the Seal are like paddles. The arm and forearm of the anterior limbs are very short, so that the paw extends but little from the body. The paw is made of what corresponds to the finger-bones in man, covered with a skin which stretches between the fingers, so as to resemble the webbed feet of swimming birds. In giving the backward stroke in swimming the fingers are spread out, but in the forward stroke they are brought together. The hinder limbs are directed backward, so as to look very much like a tail at the end of the tapering body, as seen in Fig. 45. In swimming, it uses the fore paws as paddles, and the hinder ones, with an up and down

Fig. 45.—Seal.

motion, both as a sculling and steering oar. On land or ice the movements of the Seal are very awkward, it being carried along by the fore paws, while the hinder feet are dragged along. Its body is covered with a glossy fur, closely set to the skin, so as not to interfere with its swimming, which it performs with great celerity. The nostrils and the ears have valves, which the animal can close when it goes under water, where it can, like the Whale, remain for some length of time.

102. The Seal is very useful to man. The many uses to which it is appropriated by the Greenlanders are thus spoken of by Crantz, a Danish traveler: "Its flesh supplies them with their most palatable and substantial food; the fat furnishes them with oil for lamplight, chamber and kitchen fire; and whoever sees their habitations presently finds that, even if they had a superfluity of wood, it would be of no use—they can use nothing but oil in them. They also mollify their dry food, mostly fish, with oil; and, finally, they barter it for all kinds of necessaries with the factors. They can sew better with fibres of the Seal's sinews than with thread or silk; of the skins of the entrails they make window-curtains for their tents, and shirts; part of the bladder they use as a float to their harpoons; and they make oil-flasks of the stomach. Neither is the blood wasted, but is boiled with other ingredients and eaten as soup. Of the skin of the Seal they stand in the greatest need, because they must cover with seal-skins both the large and small boats in which they travel and seek their provisions. They must also cut out of them their thongs and straps, and cover their tents with them, without which they could not subsist in summer. No man, therefore, can pass for a right Greenlander who can not catch Seals. This is the ultimate end they aspire at in all their device and labor from their childhood up."

103. Seals exist in almost every quarter of the globe, but they are mostly found in the temperate and frozen

portions, especially the latter. There are many species. The common Seal, Fig. 45 (p. 62), is from four to five feet long, and its weight is sometimes over 200 pounds. Its head is rounded, and it has long stiff whiskers. Dr. Kane's description of its appearance and habits is very graphic. In some positions it has the appearance of a dog. It has "a countenance between the Dog and the Ape—an expression so like humanity that it makes gun-murderers hesitate." It often rolls and wriggles about on the ice in the most grotesque manner, looking sometimes like an immense snail, then like a dog, and again like a couching hunter.

104. The Elephant Seal, Fig. 46, is the largest known species. It is from twenty to thirty feet long, a full grown male yielding about seventy gallons of oil. This Seal is found in the Atlantic, Pacific, and Southern Oceans. It lives in troops, migrating toward the tropics in winter, and returning toward the south pole in summer.

Fig. 46.—Elephant Seal.

It has its name on account of the long snout, which is a little like the proboscis of the Elephant, and more like that of the Tapir. When enraged, it thrusts this forward, at the same time snorting loudly. Though a formidable-looking animal, it never attacks man, but only makes a show of its large teeth to frighten him. It is sought after for its oil, and for its skin, which is much used in making stout and thick harness. The Fur Seal, found in the same quarters of the globe, has been heretofore largely taken for its skin, but it has been much thinned off, as the number taken amounted sometimes to over a million in a year.

105. The Walrus, Fig. 47, is an aberrant species. In general form and habits it is like the larger Seals. Its chief peculiarity is the great length of the canine teeth of the upper jaw, sometimes reaching to two feet. These tusks are of service in defense, in progression, and in gathering its food. It resists with them the attacks of the Polar Bear; it uses them as hooks in clambering up rocks and icebergs, and it draws up with them the seaweed which is a part of its food. It is found in the Arctic regions of both hemispheres, and is sought after for its oil and its tusks.

Fig. 47.—Walrus.

Questions.—What are included in the family of Ursidæ? Why are they called Plantigrade animals? What are Digitigrade animals? How far are the Ursidæ carnivorous? What is an omnivorous animal? What are the habits of this family? What is the type-genus of the family? How many species are there of this genus, and where are they found? What is said of their structure? What of their usefulness to man? What is said of the Grizzly Bear? What of the Polar Bear? What are some of the aberrant species of the Ursidæ? What is said of the Raccoon? Of the Badger? Of the Wolverine? Of the Kinkajou? What are the Phocidæ? Why are they called amphibions? Describe the structure and habits of Seals. What is said of their usefulness to man? Where are they found? Describe the common Seal. What is said of the Elephant Seal? Of the Fur Seal? Of the Walrus?

CHAPTER VII.

INSECT-EATING, RODENT, TOOTHLESS, AND MARSUPIAL QUADRUPEDS.

106. We now come to the second order of Quadrupeds, the Insectivora, or insect-eating Quadrupeds. Although, as we saw in Chapter III., many of the Bat and Monkey tribes live chiefly on insects, it is in this order that we find the most complete adaptation to this kind of food. The teeth of the Insectivora are not cutting and tearing, as are those of the Carnivora, but they have rounded points for the purpose of crushing the hard coverings of insects. Most of them live chiefly under ground, as the Mole; and those which inhabit cold countries are in a state of torpor through the winter. Their vocation seems to be to keep within bounds the worm and insect tribes that are found in the soil, which would otherwise be exceedingly destructive to the vegetables on which man so much depends for food.

107. Of this order there are four families: 1. Moles, which pass their whole lives in burrows. 2. Shrews, a sort of carnivorous mice, which are very common throughout Europe, but of which only a few species are found in America. 3. The Hedgehogs, found in Europe, Asia, and Africa. 4. The Banxrings, which inhabit the larger islands of the Eastern Archipelago.

108. The common European Mole, Fig. 48, lives in the same manner as the Mole of this country, although it is a differ

Fig. 48.—Mole

ent species. The eyes of the Mole are very small, as it has but little use for vision; but its hearing and smell are very acute. Its fur is fine and soft, and it will not retain a particle of dirt, although continually in contact with it. Its fore paws, mounted with strong claws, are powerful instruments for digging. In Fig. 49 you have

Fig. 49.—Fore paw of the Mole.

the bones of one of these paws, which are very large, and are worked by strong muscles. The head is constructed for digging also, the frame of the nose being wholly bone, instead of part gristle, as in most other animals. The hinder part of the body has not the great strength of the fore part, for the hind feet are not employed in digging.

109. **The plan of a mole-hill is very curious.** It has, as you see in the plan in Fig. 50, two circular galleries, one

Fig. 50.—Mole-hill.

above the other, connected together by five passages. In the very centre of the mound, and on a level with the ground around it, is a circular apartment where the Mole sleeps. This is connected by three passages with the upper gallery, and not at all with the lower one. Then there are passages running out from the lower gallery, and into one of these opens a passage from the circular chamber. Just this plan has been instinctively adopted ever since the first mole was created. The food of the Mole is chiefly worms and insects, which it gathers by burrowing. The good which the Mole does to the farmer in this way is probably much greater than any harm which his burrowing may sometimes occasion.

110. **The Shrew Mouse,** Fig. 51 (p. 68), is so called because it is so much like a Mouse, but it is readily distin-

Fig. 51.—Shrew Mouse.

guished from it by its long snout, which it uses in grubbing the earth in search of worms and insects. The Water Shrew dives and swims with great celerity, and lives on the grubs of aquatic insects, which it digs out of the mud with its snout.

111. The Hedgehog, Fig. 52, is the only animal in England that has its skin armed with spikes. These are its means of defense. When attacked, it rolls itself up, and such is the arrangement of these spikes

Fig. 52.—Hedgehog.

that the tightening of the skin makes them all stand out. A dog or a fox will not touch it then. Its food is insects, snails, frogs, snakes, roots, etc. Dr. Buckland put a hedgehog in a box with a snake. It gave the snake several quick bites in succession, rolling itself up after each bite. When the snake was sufficiently disabled, the hedgehog ate it leisurely as one would eat a radish, beginning at the tail. In winter this animal lies torpid in a hole lined with grass and moss, and if discovered looks like a ball of leaves, these having become fastened to its spikes as it rolled itself among them.

112. The Banxrings differ from the other families of this order in being arboreal in their habits, ascending trees with the agility of Squirrels, which animals they resemble in general appearance, but are easily distinguished from them by their sharp muzzles.

113. The order Rodentia, or Gnawing Quadrupeds, has eight families: 1. Squirrels. 2. Marmots. 3. Rats and

Mice. 4. Beavers. 5. Porcupines. 6. Guinea Pigs. 7. Chinchillas. 8. Hares. This order contains about three hundred species, and is the most generally distributed of all the orders of terrestrial Mammals. Its species are found in all quarters of the world, a few of them even in Australia. The furs of some of them are very valuable, as the Beavers, the Chinchillas, and the Gray Squirrels.

114. The grand peculiarity of this order is in their gnawing teeth. These are in front, two in each jaw, and they are peculiarly constructed. The front covering of the tooth is enamel, and its rear portion, that is, the body of the tooth, is ivory, which is by no means as hard as enamel. Observe the effect of this arrangement. As the upper and lower teeth are brought together in gnawing, the enamel does not wear away as fast as the ivory, because it is harder. The thin enamel, therefore, always presents a sharp chiseling edge above the level of the ivory. No other class of animals has this peculiarity. These teeth are used for different purposes, as, for example, by Squirrels in opening the shells of nuts, and by Rats in making holes in wood. The teeth of other Mammalia have a limit to their growth, but not so with these front teeth of the Rodents. These grow continually, but are kept always of the same length by the wear of the gnawing operation. If, therefore, one of them be lost, the one opposite will attain a great length. In Fig. 53 you see the lower jaw of a rabbit in which the two teeth are very long because the upper teeth were lost. A Rodent in such a plight is essentially disabled, and may die of starvation.

Fig. 53.—Overgrown Teeth of Rabbit.

115. The other teeth in the Rodents are situated far back, as seen in Fig. 54 (p. 70). These back teeth are of different kinds in the different families, according to the nature of their food. Thus in the Squirrels, which live on nuts, these teeth are rounded,

Fig. 54.—Skull of Rodent Animal.

being needed only for crushing; in the Rat they are raised into points, he being carnivorous; while in the herbivorous Rodents they are real grinders, as represented in Fig. 54.

116. The bushiness of the tail is the peculiar characteristic of the Squirrel family. This, when spread out, is of some assistance in the leaping of these arboreal animals, both guiding and buoying them up. In the Flying Squirrel, Fig. 55, there is an arrangement similar to that of the Flying Lemur, § 57.

117. The American Marmot, or Woodchuck, as it is commonly called, is about the size of a rabbit. It has an underground habitation, divided into apartments, and lives on clover and esculent vegetables.

Fig. 55.—Flying Squirrel.

Like some of the Monkeys (§ 52), it has cheek pouches, in which it carries stores of food to its burrow.

118. The Mouse and Rat family is the most numerous of all the families of the Mammalia, and contains the smallest animals. Of the common Mouse, Cuvier says, "it is known in all times and in all places." Of the Rats there are two principal species, the Black Rat, and the Brown or Brownish-gray. The Black Rat is called the old English Rat, which was introduced into England from France as late as the sixteenth century. This is now nearly exterminated by the Brown Rat, which is a stronger animal. This latter Rat was introduced into this

country at the time of the Revolution in the foreign ships.

119. The Jerboas are singular animals, making an aberrant genus of this family. They have long tails with tufted ends, and long hind legs, which enable them to make enormous leaps. The Egyptian Jerboa, Fig. 56, is about the size of a large rat.

Fig. 56.—Egyptian Jerboa.

120. Of the Beaver family, the common Beaver, Fig. 57, so well known in Canada and the northern part of the United States, is the type species. It is distinguished from all the other Rodents by its flat and scaly tail. Its hind feet are webbed, and with these and its tail it is expert in swimming. Its incisor teeth are large and uncommonly hard, and with them it can divide a

Fig. 57.—Beaver.

common walking-stick at a bite with as clean a cut as that of a hatchet. Like the Seal (§ 101), it can close its ears and nostrils when it dives into the water. Beavers are very celebrated for the skill with which they build their dams and habitations, which they always do in companies.

121. The common Porcupine, Fig. 58 (p. 72), is found

Fig. 58.—Porcupine.

in Africa, India, Persia, Tartary, and in some parts of Europe. It is nearly the largest of the Rodents. The spikes or quills with which it is covered constitute, like those of the Hedgehog, its means of defense. If it can not escape, it stands still, with its quills all bristling, or even runs back against its adversary. The fact that any quills that are a little loose fall off, or remain sticking to an adversary, has given rise to the mistake that the animal has the power of shooting them from its body.

122. Of the Guinea-pig family, the Capybara, Fig. 59,

Fig. 59.—Capybara.

is the largest of all the Rodents. It is a native of South America, where its flesh is much prized. It is a favorite prey of the Jaguar. Its shape, and its thin and straight hair, make it look quite like a pig.

123. The Hares differ from the other Rodents in having more than four front sharp teeth. There are about thirty species. The Hare, which in England furnishes

such rare sport to the hunters, is represented in Fig. 60.

Fig. 60.—Hare.

The Rabbit, which is every where domesticated, is smaller than the Hare, but is like it in form. It lives in a burrow, while the Hare lives in a sort of nest which it constructs from grass.

124. We now pass to the Edentata or *toothless* Quadrupeds. This term applies only to a part of the order, the Ant-eaters and the Pangolins. The Sloth and the Armadilloes have back teeth, but they are imperfect.

125. That singular animal, the Crested Ant-eater, Fig. 61 (p. 74), is found in Guiana, Brazil, and Paraguay. It is nearly four feet long. It lives both on common ants and the termites or white ants. With its strong claws it tears open their habitations, and then thrusts in its long tongue. This, being covered with a gummy saliva, has, when withdrawn, a multitude of ants adhering to it, which the animal swallows.

126. The Pangolins, or Manidæ (plural of Manis), are ant-eaters, and take the ants in the same way that the Crested Ant-eater does. They are remarkable for being encased in an armor of horny scales. When attacked, they roll themselves up, and raise their sharp-edged scales

Fig. 61.—Crested Ant-eater.

as the Hedgehog does his spines. The Long-tailed Manis, Fig. 62, is a native of Africa.

Fig. 62.—Long-tailed Manis.

127. The Armadilloes are found only in South America. The armor which covers them is different from that of the Pangolins. It is a sort of plate-armor. One species, the Six-banded, is represented in Fig. 63 (p. 75). The natives consider these animals a great delicacy when

TOOTHLESS QUADRUPEDS. 75

Fig. 63.—Six-banded Armadillo.

roasted in their shells. The Armadilloes live on carrion, insects, and fruit. They are all small, except one species, which is called the Gigantic Armadillo, and weighs a hundred pounds or more.

128. The Sloth, Fig. 64, differs from all other arboreal

Fig. 64.—Sloth.

Quadrupeds in its manner of climbing. It always has its back downward, as seen in the figure. It has been common to consider this animal as imperfectly constructed, and even Cuvier speaks of the "inconveniency of its organization," and says of it that "nature seems to have amused herself in producing something grotesque and imperfect." But there is perfect adaptation here, as in every other animal, of the organization to the habits. It is constructed to live just in the way that it does, and moves about in the trees with great facility. It has been known to go from the bottom to the top of a high tree in a minute's time. With its strong curved claws it sleeps hanging from the branches of a tree as easily as a bird sleeps on its perch. The three species of Sloths are found only in the forests of the tropical portion of South America. They live on the leaves of trees.

129. The order of Marsupials is named from a pouch or bag (Latin, *marsupium*) which the females have for carrying their young for some time after birth. The young are born in an immature helpless state, and a sort of *nest* is thus provided for them in the body of the mother. Even after they have become able to leave it, they flee to it whenever they are alarmed. There are about eighty species. All of these animals are found only in Australia and the neighboring islands, except the Opossums, which are found on the western continent, especially in South America.

130. The Great Kangaroo, Fig. 65 (p. 77), a native of Australia, has very long and powerful hind legs, and can make leaps of fifteen feet. Its fore feet are short and small, and are used more as hands than as feet. Its length is about five feet, and its tail is three feet long. There are many different species of Kangaroos, all having a general resemblance to this.

131. The Opossums are peculiar to America. There are about twenty species. They are arboreal in their habits, and they are assisted in their climbing, like some of the Monkey tribe, by their tails, which are long and scaly. In one other respect they are still more allied to the Monkeys. The inner toe of the hinder foot is some-

Fig. 65.—Great Kangaroo.

what like a thumb, as it can be brought in opposition to the other toes for grasping. They can therefore be called, like the monkeys, Pedimana, or foot-handed animals. The pouch in the abdomen for their young, however, places them decidedly among the Marsupials.

Fig. 66.—Virginia Opossum.

132. The Virginia Opossum, Figure 66, found in many of the Southern States of this country, is one of the largest of the genus, being about the size of a cat. It is nocturnal and arboreal. It remains in

the daytime inert in branches and hollows of trees, but prowls at night in search of its food, which consists of insects, birds, eggs, fruits, etc. It makes great use of its tail in climbing, being able to swing by it from one branch to another. When attacked it feigns death, and so well that even dogs are deceived. This is the origin of the common phrase, " playing 'possum."

133. There are two very singular animals in Australia, about the classification of which there has been some difference of opinion. By some they have been placed in this order, on account of some resemblance in the skeleton, although they have not any marsupium. The first is the Duck-billed Platypus, Fig. 67. This singular ani-

Fig. 67.—Duck-billed Platypus.

mal has a body like that of an Otter, and a bill like that of a Duck. It was first made known to British naturalists by a stuffed specimen, and it was at once suspected that the bill of some Australian bird had been ingeniously fastened to the head of a quadruped. But it was found to be no deception, and this animal presents the strongest example that we have of an approach in the Mammal tribe to that of birds. It uses its bill precisely as the Duck does, searching for insects, small shell-fish, etc., by plunging it here and there in the mud. There is a curious provision in the young to prevent the bill from interfering with the operation of suckling. It is very soft, and does not become hard till it is time for the animal to cease to suckle. The fore feet are formed for digging, and the animal excavates a burrow, sometimes even fifty feet in length, in the bank of the stream, where it lives. Both the fore and the hind feet are fitted for swimming by being webbed. The web on its fore feet extends over its claws, but it has the power of folding it back when it wishes to dig.

134. The other animal is the Echidna, or Porcupine Ant-eater. It is about the size and form of a hedgehog, but its spines are stouter. It burrows with great rapidity. When attacked by dogs, it quickly, by digging, sinks itself in earth or sand, so that they can see nothing but its bristling back, and this they are not disposed to touch.

Questions.—What is said of the structure and habits of the Insectivora? What are the families of this order? What is said of the structure and habits of the common Mole? Describe the arrangement of a mole-hill. What is said of the Shrew Mouse? What of the Hedgehog? Of the Banxrings? What are the families of the order Rodentia? Describe their front teeth. What is said of their back teeth? What is said of the Squirrel family? What of the American Marmot? Of the Rats and Mice? Of the Jerbons? Describe the structure and habits of the Beaver? What is said of the Porcupine? Of the Guinea-pig family? Of the Hares? What is said of the Edentata? What of the Crested Ant-eater? Of the Pangolins? Of the Armadilloes? Of the Sloth? From what do tne Marsupials get their name? Where are they found? What is said of the Great Kangaroo? How many species are there of the Opossums? Where are they found? What are their habits? How are they allied to the Monkey tribe? What is said of the Virginia Opossum? What are the structure and habits of the Duck-billed Platypus? What is said of the Echidna? Where are these two animals found?

CHAPTER VIII.

THICK-SKINNED QUADRUPEDS.

135. We now come to the second division of Quadrupeds, the Ungulata or hoofed quadrupeds. Of this there are two orders: 1. The Pachydermata, or thick-skinned. 2. The Ruminantia, ruminating, or cud-chewing. The Pachydermata are variously classified by different naturalists. I make six families: 1. The Elephants. 2. The Tapirs. 3. The Pig Family. 4. The Rhinoceros Family. 5. The Hippopotamus Family. 6. The Horse Family.

136. Of the Elephants there are only two species, the Asiatic and the African, the latter of which you see in Fig. 68. The Elephant has several hoofs arranged in a

Fig. 68.—The Elephant.

circular manner around the bottom of the foot. His trunk or proboscis (from which this family is sometimes called Proboscidea) is a wonderful organ. It has in it 40,000 muscles interlaced together. These give it great flexibility, and make it the hand of the elephant. On the end of this hand is a small finger-like projection, which is a feeler, and is also used in picking up small objects. The Elephant gathers his food with his trunk, and puts it into his mouth. He gets his drink also with his trunk in this way—he draws it up into the two nostrils of the trunk, it being prevented from going back into the throat by a valve. When he drinks he turns the end of the trunk into his mouth, and pours the water in from it. He sometimes gives himself a shower-bath by throwing water from his trunk over his body. It is through the trunk that the Elephant sends forth his trumpet-like voice. This organ is not only a hand, a forcing and suction pump, and a trumpet, but it is also the animal's nose.

137. The neck of the Elephant is so short that he could not possibly reach his food or drink without his trunk. His food is chiefly grass, the leaves of trees, and roots. These last he loosens with his tusks, using them as we use a crowbar, and then he pulls them up with his trunk.

138. Elephants congregate in large herds, sometimes numbering hundreds, or even thousands; and no sight can be more grand than such a herd in the midst of the magnificent scenery and rich verdure of an African landscape. The Elephant of India is more sagacious than that of Africa, and is much used in traveling, and in hunting tigers, as described in § 72. The African Elephant is not at present tamed by man, and is hunted merely for the sake of his tusks, from which very fine ivory is obtained. The trade in tusks, both in Asia and Africa, is immense. It requires annually many thousands of elephants to furnish a supply of ivory to England alone.

139. Although the Elephant is the largest of all the terrestrial Mammalia, there are remains of extinct animals which reached a much larger size. This is the case

with the Mastodon Giganteus, whose bones have been found alone in America.

140. The Tapir is in some respects like the Hog. It has a prolonged snout, which allies it, on the other hand, to the Elephant. With this it grasps fruit and herbage, putting it into its mouth. The South American Tapir is from five to six feet high. The Malay Tapir, Fig. 69,

Fig. 69.—The Tapir.

is larger. It has its loins and hind quarters of a grayish white color, giving it a singular appearance.

141. Of the Pig Family I need say but little. The two orifices of the snout are like those in the trunk of the Elephant. The proverbial uncleanliness of the common Hog is owing in fact to the circumstances in which man places it, and no animal seems to like clean straw better. The Wild Hog or boar, the original of the domestic hog, is still found in many parts of Europe, especially in the German forests, and its chase is one of the sports of hunters. One of this family, the Babyroussa, or hog-deer, Fig. 70 (p. 83), has four tusks, two of which do not pass out between the lips, but through an opening in the skin. It is a native of Java and the Moluccas.

THICK-SKINNED QUADRUPEDS. 83

Fig. 70.—The Babyroussa.

142. There are seven species of the Rhinoceros. These are ungainly animals with short legs, approaching in size the Elephants. They are distinguished chiefly by their horns, which are in texture something like whalebone. Some species have two horns. Those that have one, as in Fig. 71, are called unicorns. These animals live an

Fig. 71.—The Rhinoceros.

indolent life on the marshy borders of lakes and rivers, and are very fond of wallowing in mud. They are found in Asia and Africa.

143. There is but one known species of the Hippopotamus (ἱππος, *hippos*, horse; ποταμος, *potamos*, river), or river-horse, an inhabitant of Africa. It passes a large portion of its time in the water, especially in the daytime, leaving it at night in search of its food, which is the herbage growing on the banks of rivers and lakes. Its hide is of great thickness, even to two inches, on its back and sides, and is made into shields, whips, and walking-sticks. This animal is supposed by some to be the Behemoth of the Bible.

144. There are certain birds, called Rhinoceros Birds, which are always in attendance on the hippopotamus and the rhinoceros. They live on the ticks and other parasites which swarm upon these animals. It is said that these birds are the best friends which those huge creatures have, for they rouse them from their sleep when they see an enemy approach.

145. The Horse Family includes the Horse, the Ass, the Zebras, etc. The hoof in this family is one solid piece, and so the family is sometimes called solidungula.

146. The first mention made of the Horse in the Bible is in connection with the sale of corn in Egypt by Joseph, Genesis, xlvii., 17. What is the original country of the horse is not known. The herds running wild in Tartary, Carpenter says, are undoubtedly descendants of horses that have been domesticated, for their habits are the same with those of the herds in the pampas of South America, and these are known to have descended from horses introduced by the Spaniards. The herd has always a leader which is a male, and when attacked they put the colts and the females in the rear, and make resistance by kicking with their hind feet. The natives catch these wild horses with the lasso, a noose of leather, which they throw with great skill, and they very readily tame them. There are herds of wild oxen as well as horses in the pampas of South America, and there is accordingly an immense trade in the hides of both.

147. The finest horses in the world are found in Arabia, and nowhere is this animal more highly prized. The Arab treats his horse as one of the family, permitting him to live in the same tent with him, to feed from his hand, and even to sleep among his children. The mutual attachment between the horse and his master is therefore often of the strongest character, and the most extravagant offers will sometimes fail to induce an Arab to part with his horse, even when pinching poverty makes these offers very tempting to him.

148. The Ass was domesticated probably before the Horse. It was, and is now, in many parts of the East, the beast usually ridden in civil life, the Horse being especially devoted to war. The care bestowed upon it there makes it really an elegant and spirited animal. The custom of having persons of distinction ride on white asses is of great antiquity, as appears from Judges v., 10, "Speak, ye that ride on white asses." Some asses are fleeter than the Horse, as the Dzigguetai, Fig. 72, which inhabits the greater part of Central Asia.

Fig. 72.—The Dzigguetai.

149. The Zebras, Fig. 73 (p. 86), found in Southern Africa, live, like the horse, in troops, and, with their dis-

Fig. 73.—The Zebra.

tinct and regular stripes, make a brilliant appearance as they flee together before the hunter. The Quagga of the same country is similar, but from the indistinctness of its stripes it is a less beautiful animal. Neither of these animals can be profitably used like the Horse and Ass, because they are so wild and vicious.

Questions.—What are the orders of the Unguiata? What are the families of the Pachydermata? How many species are there of the Elephant, and where are they found? What is the arrangement of the Elephant's foot? What are the various offices of his proboscis? What is his food, and how does he obtain it? What is said of the herds of elephants? How are Elephants valuable to man? Which species is most so? What is said of large extinct animals? What is said of the Tapirs? What of the Pig Family? What of the Babyroussa? Of the Rhinoceros? Of the Hippopotamus? Of the Rhinoceros Birds? What does the Horse Family include? What is the first that we know of the Horse? What is said of the herds of wild horses? What of the Arabian Horse? What of the Ass? What species of the Ass is fleeter than the Horse? What is said of the Zebra?

CHAPTER IX.

RUMINANT QUADRUPEDS.

150. Of the Ruminantia, or cud-chewing quadrupeds, there are eight families: 1. Bovidæ, oxen, buffaloes, etc. 2. Ovidæ, sheep. 3. Capridæ, goats. 4. Cervidæ, the deer tribe. 5. Moschidæ, the musk-deer tribe. 6. Antelopidæ, antelopes. 7. Camelidæ, camels. 8. Camelopardæ, giraffes, or camelopards.

151. No animals are so useful to man as those of this order. Almost all the animal flesh which he consumes comes from the Ruminants. Some of them are his beasts of burden, and some supply him with various articles of necessity and convenience, such as milk, tallow, hides, horns, etc. Being thus necessary to man, they are distributed over nearly all parts of the globe. Some of them, as the Reindeer of Lapland, and the Camel of Arabia and Northern Africa, are confined mostly to certain regions; while others, as the Ox, the Sheep, and the Goat, go every where with man, except in regions which are so cold as not to afford them the requisite food in pasturage.

152. The Ruminants make a very well defined order, all the families agreeing in their prominent common characteristics, and none of them being to any extent aberrant. Of all the herbivorous animals these are the most entirely confined to vegetable food. Of the Rodents, though mostly herbivorous, there are many that eat some animal food; most of the Edentata live on insects, and some devour flesh; and several species even of the Pachydermata have in part an animal diet. But there is not one of the Ruminants that is not exclusively herbivorous. Some, as the Camel and the Giraffe, are formed for browsing on the leaves and young shoots of

trees, but most of the order are fitted to gather and live upon the herbage on the surface of the ground.

153. The feet in this order agree in terminating in two toes with hoofs. These appear externally as if there was a single hoof cleft. Hence these animals have been called cloven-footed. No animal in this order has front or incisor teeth in the upper jaw. There is a firm pad there, against which the incisor or cutting teeth of the lower jaw press when the jaws are brought together. The back teeth are specially formed for grinding, and the jaws are adapted to the sidewise grinding motion. The difference between this and the motion of the jaws in a carnivorous animal, you may see if you observe a cow and a dog when eating.

154. The name of this order is given to it from the singular process called *rumination*. The object of this I will explain. The stomach of the Ruminant is not a single organ. It has four cavities, as you may see in Fig. 74, in the case of the Sheep; or, rather, there are four

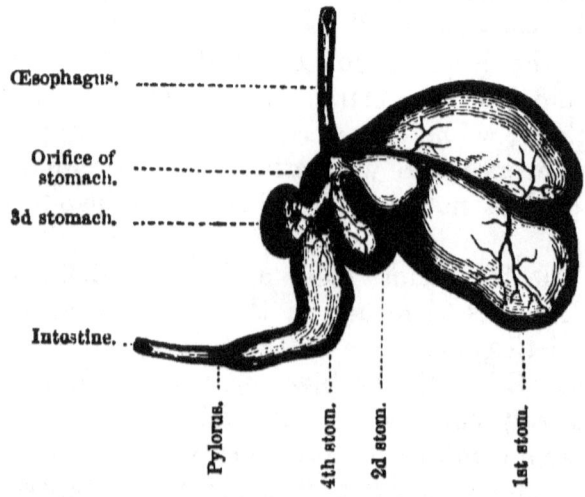

Fig. 74.—Stomachs of the Sheep.

stomachs. The grass cropped by the Ruminant animal is not chewed at once, but is passed directly into the large first stomach, or paunch. Here it is macerated or soaked. Then it is passed into the second stomach, or honeycomb stomach, as it is called, from the cellular arrangement of its inner surface. Here in some way it is all made into distinct balls. Each of these is passed up into the mouth, and is chewed. It then goes down the gullet into the third stomach, the *manyplies*, so called because its inner lining membrane has a great many folds. From thence it is passed into the fourth stomach. It is this that corresponds to the stomach of man, and of all animals that live partly or wholly on animal food; for here the gastric juice is secreted and is mingled with the food. In the suckling Ruminant the milk passes directly into the fourth stomach, the other stomachs remaining unemployed until the animal begins to graze.

155. The purpose of this arrangement for rumination is thus stated by Carpenter: "The Ruminantia, taken as a group, are timid, and are destitute of powerful means of defense against their foes, seeking safety in flight when alarmed, rather than stopping to defend themselves. A large proportion of them are natives of tropical regions, where they are liable to the attacks of the larger beasts of prey. Now their food—consisting, as it does, of grasses and herbage, which contain a considerable amount of woody fibre—requires to be thoroughly masticated before it can be properly digested. When feeding on the pastures they frequent they are subject to many alarms; and if they were compelled to spend a considerable time in masticating their food before swallowing it, they would often be in danger of starvation, by being obliged to leave their pasture before their wants were supplied. But by their power of subsequently returning their food to the mouth, and chewing it at their leisure, they are enabled to dispense entirely with any mastication previously to first swallowing it, and to feed with comparative quickness. They thus convey a store of food into the first stomach or paunch, as the Monkey does into his cheek-

pouches; and then, retiring to a secluded place among their mountain fastnesses, they masticate their aliment in comparative security. Moreover, the maceration (or soaking) in the fluids of the first and second stomachs, to which the food has been subjected, causes it to be much more readily ground down than if it were triturated immediately on being first cropped from the pasture."

156. There is an obvious adaptation of the structure of the Ruminants to the habits just stated. That they may quickly perceive the approach of an enemy their senses are extremely acute. Their eyes are placed at the side of the head rather than in front, which affords them a great range of vision. Besides this, the pupils of the eyes have an oval shape, extending *horizontally*, instead of up and down, as we see it in the Cat.* This increases the range of sight in the rear direction. The ears are placed far back, and can be readily turned to any quarter. This is quite essential in fleeing from their pursuers. In order that they may flee swiftly they have long legs, and are for the most part slender in form. When there is an accumulation of flesh and fat, making the animal bulky and slow in motion, it is commonly owing to the influence of domestication. Though the Ruminants are generally timid animals, the means of defense which they have in their horns and hoofs some of them are disposed to use sometimes in offensive warfare, at least among each other.

157. The family Bovidæ (*Bos*, an Ox) is distinguished from the other families of this order by the uniform presence of horns in both sexes, and by the bulkiness of their forms. The common Ox is diffused widely in all quarters of the globe, and has a great variety of breeds. I will notice only one. The Bos Indicus, the Zebu, or

---

* The reason for this shape of the pupil in the cat and other animals of the feline tribe is obvious. In taking its prey the animal has need of a good range of vision up and down, or vertically, rather than laterally, especially if its prey be on any height, as a tree.

Brahmin Bull, Fig. 75, is a native of India, and is remarkable for a large fatty hump above the shoulders. In all

Fig. 75 —Zebu, or Brahmin Bull.

Southern Asia and Eastern Africa this animal supplies the place of the common Ox, and is supposed to have come from the same origin, instead of being another species. The Hindoos treat it with great reverence and attention. They allow it to go about the streets, which it does with great familiarity, even walking into shops, helping itself to sweetmeats and other articles, and resenting the slightest affronts with a peevish thrust of the horns. But while the bull is thus honored, the ox is treated without mercy, being urged on in its labor by the cruel goad. The Brahmin cow is treated more kindly than the ox, but is not reverenced as the bull is.

158. The true Buffaloes belong to a genus of this family. They are found in Asia and Africa, and to some little extent in the south of Europe. The common species, Fig. 76 (p. 92), was originally a native of India, where it has long been domesticated, and used like the Ox. Its hide is very strong, and harness is made from it.

159. The American Bison, Fig. 77 (p. 92), improperly called a Buffalo, is found in immense herds in the prairies

92   NATURAL HISTORY.

Fig. 76.—Buffalo.

of North America. The Indians hunt them with the bow and arrow, mounted upon swift horses to give them chase.

Fig. 77.—American Bison.

They show great skill as well as daring, often firing their arrows into the hearts of their victims. The flesh of these animals constitutes a large portion of the food of the Indians. Much of the *pemmican*, so called, used by hunters and voyagers in the far north, is made from the meat of the Bison. Then the skin, the buffalo-robe, is a necessary article of clothing, and is used also in constructing tents, and the horns furnish the powder-flasks of the hunters. The Buffalo or Bison hunt is therefore a great item in the life of an Indian in the West. The herds of these animals sometimes number thousands. Lewis and Clarke supposed that there were certainly 20,000 in one herd which they saw. The range of the Bison in this country is becoming every year less extensive from the encroachments of civilized man.

160. The Yak, Fig. 78, is found in Tartary. It is not

Fig. 78.—Yak.

a very large animal. The mass of hair, which, rising above the shoulders, hangs like a mane almost to the ground, is applied to various uses by the Tartars. They weave it into cloth, which they use in making articles of dress and their tents, and they also make ropes from it. The hair of the tail, which is great in amount, is long and fine. The tail, with an ivory or metal handle, is used in India to keep off musquitoes, and is called a chowrie.

161. The Musk Ox, Fig. 79, is a native of the cold regions of North America. It somewhat resembles the

Fig. 79.—Musk Ox.

Yak. It is covered with very long hair which almost reaches the ground. It appears in small herds, numbering, perhaps, twenty or thirty. Both this animal and the Yak are rather small, but the thick hair covering them makes them look quite large.

Questions.—What are the families of the order Ruminantia? What is said of the usefulness of the Ruminants to man? State how well defined this order is compared with some others. What is the structure of the feet of the Ruminants? What are the structure and arrangement of their teeth? What is rumination? Describe the arrangement of the stomachs of the Ruminants. Illustrate its purpose. What is there in some Monkeys analogous to the paunch of the Ruminants? In what other respects is the organization of the Ruminants adapted to their habits? What is said of the arrangement of the eye? What influence has domestication on the bulk of the Ruminants? What partial exceptions are there to the general timid habits of this order? What distinguishes the Bovidæ from the other families? What is said of the distribution of the Ox, and of its varieties? What is said of the Bos Indicus? Where are the true Buffaloes found? How are they useful to man? What is said of the American Bison? What of its usefulness to man? What is said of the Yak? What of the Musk Ox?

CHAPTER X.

RUMINANT QUADRUPEDS—*continued.*

162. THE different species of the Ovidæ, or Sheep family, have many varieties, from the influence of domestication. The Sheep is the first animal noticed in the Bible as subjected to man, for "Abel was a keeper of sheep." The tail of the Sheep seems to be much affected by domestication, it being much larger in the domesticated than in those that run wild. In the Egyptian and Syrian Sheep it often becomes enormous, reaching a weight of 50 or even 100 pounds, in which case a board or a little wagon is attached to it, to prevent it from dragging on the ground. This overgrown tail is mostly a mass of fat, which is considered a great delicacy, and is frequently used as butter.

163. The Capridæ, or Goat family, are nearly allied to the Sheep. They are, however, stronger, lighter, more agile, and less timid. They appear in almost all parts of the world. In some countries they are greatly valued for their milk. The best Morocco leather is made from their skins, and the skin of the kid is much used in making fine gloves. The silken wool of the Angora Goat of Asia Minor hangs in long ringlets, furnishing the material for the finest camlets. From the wool of the Cashmere Goat of Thibet and the region of the Himalaya Mountains, are manufactured the famous Cashmere shawls. The Caucasian Ibex, Fig. 80 (p. 96), which inhabits the Alpine regions of Europe and Western Asia, is remarkable for its large and beautiful horns. They are surrounded with rings at regular intervals, and are very strong. When chased, it will frequently turn on its pursuer, and with its horns, hurl him from some

Fig. 80.—Caucasian Ibex.

precipice, unless he can shoot it before it reaches him.

164. The Cervidæ, or Deer family, are distinguished from all the other families of Ruminants, in having horns which are cast off at intervals, new ones growing out in their place. In the young animal they are small, but in the full-grown Deer they are very large. These horns are also covered with a velvety skin, and are called antlers. While they are growing there are blood-vessels in this skin, and from the blood in them the antlers are made. You can see on them, after this skin is stripped off, just the course of the large arteries, by the channels for them in the horn. These antlers grow very rapidly. After they have attained their growth, there is no farther need of the blood in the "velvet," and it must be got rid of, for if it remained there would be bleeding every time that the Deer should hit any thing hard with its antlers. There is a singular process for doing this. In the rings of bone at the foot of the antlers there are openings, through which the arteries pass. These gradually close up, and the supply of blood to the "velvet" is, therefore, gradually cut off. It would not answer to have this done suddenly, for then all the blood going to the head would be turned in upon the brain, and such a rush of blood to that organ would be injurious, perhaps fatal. After blood ceases to be supplied to this skin it dries and readily peels

off, and the Deer gets rid of it by rubbing his antlers against the trees.

165. The females of this family, except in the case of the Reindeer, have no antlers. In those species that are found in extremely cold climates, as the Elk, Fig. 81,

Fig. 81.—The Elk.

the antlers are apt to be flattened, "as if," says Carpenter, "they were destined to be used by the animal, like shovels, in clearing the snow from off its food." The animals of this tribe are celebrated for both their beauty and speed. They are distributed over all parts of the globe, except Australia, and the southern and central regions of Africa, these regions being supplied in place of them with Giraffes and multitudes of Antelopes.

166. The Reindeer is seen throughout the Arctic regions of America, Europe, and Asia. It lives in summer on the buds and twigs of small shrubs, and in winter on a lichen growing under the snow, which it digs up with

its feet. It is gregarious both in the wild and in the domesticated state. So important is this animal to the Laplander, that his wealth is estimated by the number of Reindeer which he has, just as that of the patriarchs of old, and the Arabs of the present time, is estimated by the number of their herds, and flocks, and camels. A Laplander in good circumstances has several hundred, and some have not less than two thousand. The Gadfly and the Mosquito are so annoying to the Reindeer, that the Laplander is obliged to make periodical migrations with his herd to the mountains to escape them.

167. The Axis Deer, Fig. 82, is a beautiful animal. It

Fig. 82.—The Axis Deer.

is a native of India. Its horns are slender, and are divided quite regularly into three branches. Its usual color is a fawn yellow, with regular white spots, and a black stripe running down the back.

168. The Moschidæ take their name from that peculiarly strong perfume called musk, which is obtained from one of the species. They resemble the Deer family in

general appearance, but they are much smaller, and they have no horns. The true Musk-deer, Fig. 83, is found

Fig. 83.—The Musk-deer.

in the central part of Asia. The musk is contained in a pouch. Its perfume is so strong when pure and fresh, that the hunter, after killing the animal, is obliged to cover his nostrils with cloth before he secures the pouch, else he will have severe headache, and perhaps violent bleeding from the nose.

169. This substance, the most powerful perfume in the world, is formed from the blood of the animal, like any other secretion. And yet his blood does not differ essentially from that of other animals, neither is his food especially different from that of those in the same neighborhood. The chemistry which can produce this, and various other perfumes in other animals, is utterly beyond our knowledge. The same thing can be said of the poisons in both the animal and vegetable world, they being made in the animal from the blood, and in the vegetable from the sap.

170. The Antelopes are similar to the Deer in general form and in activity. They differ from them chiefly in having permanent horns. There are more than seventy species distributed through the warm parts of the earth. They are most abundant in Africa, a few species being found in Asia, fewer still in America, and only two in

Europe. They may be divided into four sub-families:
1. The true Antelopes, remarkable for their graceful forms, long and slender limbs, and great agility. 2. The Bush Antelopes, having a more compact form and shorter limbs, and living in jungles and thickets. 3. The Capriform Antelopes, shaped much like goats, and living on hills and mountains; the Chamois of Europe is of this kind. 4. Bovine Antelopes, verging in their shape to the Ox family; this may be considered as a decidedly aberrant group. I will notice but a few of the species of the Antelope tribe.

171. The Springbok, Fig. 84, is one of the most beau-

Fig. 84.—The Springbok.

tiful and agile of the true Antelopes. It inhabits southern Africa. It derives its name from the habit which it has of springing up to the height of several feet when alarmed. Large herds of Springboks spread themselves over the wide plains. When a drought occurs, as is often the case in the tropical regions, they migrate in large bodies in search of food. Some persons have seen, as they suppose, as many as twenty or thirty thousand together.

172. Among the true Antelopes is also the Gazelle,

Fig. 85.—The Gazelle.

Fig. 85, so celebrated in the poetry of the East. This is probably the Roe of the Bible. Its eyes are large, dark, and lustrous. Its speed is so great that not even

Fig. 86.—The Oryx.

the Greyhound can overtake it. It lives in herds, and is found in Arabia and Syria. It is easily domesticated, and is often seen in the court-yards of houses in Syria.

173. The Oryx, Fig. 86 (p. 101), is a native of South Africa. It is the swiftest of all animals in that region. It has many of the characteristic beauties of the Antelopes, but its tail is like that of a horse, and its horns are very peculiar, being perfectly straight and of a dark color. With these formidable horns, two and a half feet in length, it can defend itself even against the Lion. When the Lion attacks it, it lowers its horns and receives him on its sharp points; and the two have been often known to die together, the Oryx by the violence of the shock and the Lion from the wounds of the horns.

174. The Kudu, Fig. 87, also a native of South Africa,

Fig. 87.—The Kudu.

is one of the most beautiful of the Antelopes. Its horns are nearly four feet long, and their spiral form adds much to their beauty. Although a large animal, it can leap with wonderful activity. The largest of the Antelopes is the Eland, found in the same region. It is as large as an ox. It is hunted for its flesh, which is highly esteemed.

175. The Gnu, or Horned Horse, Fig. 88, is a very

Fig. 88.—The Gnu.

singular animal belonging to this same region. It is difficult, at first view, to say whether it has most of the characteristics of the Horse, or the Buffalo, or the Antelope. Its horns cover the top of the forehead, then, sweeping down in front of the face, turn with a sharp curve upward. This is like some of the Buffaloes. The resemblance to the Horse is in the mane and the tail. The legs are like those of the Antelopes. It is an animal of great speed. When enraged it is very dangerous.

176. The family Camelidæ includes the Camels and Dromedaries of the Old World, and the Llamas, which may be said to be the Camels of the New. There are two species of the true Camel: the Arabian Camel, Fig. 89 (p. 104), having one hump, and the Bactrian Camel, having two humps, the latter being an inhabitant of Central Asia, Thibet, and China.

Fig. 89.—The Arabian Camel.

177. The Arabian Camel has been called very appropriately "the ship of the desert." It is especially fitted in many respects for traveling across the wide deserts in that quarter of the world. Its broad elastic cushions on its feet afford it a firm footing on the sand. The callous surfaces on its chest and limbs defend it from the heat of the sand as it takes its rest. The eye is shielded from the glaring light of the sun by a brow hanging over like a roof, and by its long eyelashes. Its nostrils can be closed at pleasure when the hot sand is driven along in clouds by the wind. Its teeth and lips are fitted to the food on which it must depend in the desert. The thorny shrubs and tough leaves which it eats require powerful cutting and grinding teeth for their mastication. These the Camel has. And with its long stout lip it readily draws the twigs and leaves into its mouth. But the most essential provision of all is in the water-cells in one of the stomachs of the Camel. Here he can stow away a large quantity of water for use on his long journey. This he uses only as he requires it. When

he is thirsty, or needs water to moisten his food as he eats it, he can force any amount that is required out of this reservoir up into the throat. By this arrangement the Camel can go without drinking for many days. Sometimes travelers, who are suffering severely from want of water, kill one of the Camels in their caravan for the purpose of getting at the water in this reservoir.

178. The Camel is a strange-looking animal. The Pictorial Museum contains the following good description of it: "There is something strange and imposing in the aspect of the gaunt and angular Camel, destitute, as it confessedly is, of grace and animation. We are amazed at its height, its uncouth proportions, its long, thin neck, its meagre limbs, and the huge hump on its back, which conveys the idea of distortion. Quietly it stands in one fixed attitude, its long-lashed eyelids drooping over the large dark eyes; it moves, and onward stalks with slow and measured steps, as if exercise were painful. To complete the picture, it is covered with shaggy hair irregularly disposed, here forming tangled masses, there almost wanting. Its thick mobile upper lip is deeply divided; its feet are large and spreading, the toes being merely tipped with little hoofs."

179. The docility of the Camel is such that one man can lead thirty, or even fifty of them, fastened together in a row. The traveler mounts the Camel as it is kneeling; and as it rises, unlike the well-known habit of the horse, upon its hind feet first, he will be thrown suddenly over its head unless he is especially careful. The importance of the Camel in the regions where it is found can hardly be realized by us. It is essential, as you have seen, wherever wide deserts are to be traversed; and St. Hilaire, in his Letters on Egypt, says that "without it nearly the whole of Africa and one quarter of Asia might perhaps have remained uninhabited." This statement is rather too strong, but it shows what is the estimate of the Camel's value by one who had traveled extensively

in those regions. Besides its uses as a beast of burden, this animal affords sustenance to man by its milk and its flesh, and also hair for the manufacture of cloth.

180. The Dromedary is a mere variety of the Camel, holding the same relation to it as a race-horse does to the heavy draft-horse. It is used principally for journeys where dispatch is requisite; and it can carry only a single person, and but a light burden in addition. It is by no means as fleet as a horse, but it can maintain a moderate pace for a long time, going easily at the rate of six or even eight miles an hour for twenty-four hours consecutively.

181. The Llamas of South America, of which there are several species, though they are much smaller animals than the Camel, resemble it in many respects in form and structure. They have, however, no hump, and their feet, instead of being cushioned, have hoofs with claw-like projections, to enable them to climb the rocky hills among which they live. The Peruvian Llama, Fig. 90, inhabits

Fig. 90.—Peruvian Llama.

elevated regions, almost on the borders of perpetual snow. When the Spaniards came first to South America, this animal was the only beast of burden; but now it is superseded mostly by the horses and mules which have been introduced from Europe.

182. In the family of Camelopardidæ there is only one known species, the Giraffe, Fig. 91. This very peculiar

Fig. 91.—Giraffe.

animal has some points of resemblance to the Camel, and some to the Deers and Antelopes. It is found only in Africa; there being two varieties, one in the southern part of the continent, and the other in Nubia, Abyssinia, and the adjoining districts. It is seen in herds of twelve to forty in number, making splendid objects in the landscape, as with their tall necks they browse from the trees.

Questions.—What is said of the Ovidæ? What effect is produced on their tails by domestication? How do the Capridæ differ from the Ovidæ? What is said of their usefulness to man? What is said of the Angora Goat? Of the Cashmere Goat? Of the Caucasian Ibex? How are the Cervidæ distinguished from the other families of the Ruminants? What is the office of the "velvet," and how is it disposed of when no longer needed? What is said of the Elk? Of the Rein deer? Of the Axis Deer? How do the Moschidæ differ from the Deer family? What gives them their name? How is the musk obtained? What is said of the chemistry of this secretion? Compare the Antelopes with the Deer. What countries do they inhabit? Into what sub-families are they divided? What is said of the Springbok? Of the Gazelle? Of the Oryx? Of the Kudu? Of the Gnu? What are included in the family Camelidæ? Where are the two species of Camels found, and how do they differ? Show in what respects the organization of the Camel is adapted to its habits and circumstances. Describe its appearance. What is said of its docility? Of its mode of rising from a kneeling posture? Of its importance to man? What is said of the Dromedary? What of the Llamas? What of the Giraffe?

CHAPTER XI.

THE WHALE TRIBE.

183. THE water contains both the largest and smallest of animals. In the sub-class now to be considered, the Cetacea, or Whale tribe, we find the largest animals existing at the present time. Those monstrous terrestrial quadrupeds, the Elephant and the Hippopotamus, are not to be compared to the Whale; and even the smaller species of this class, the Dolphin and Porpoise, are above the average size of land animals.

184. The animals of this tribe are, unlike all that we have as yet considered, destitute of both hands and feet. Though they are Mammals, they are fitted to live, like the Fishes, in the water. They were classified among fishes by ancient zoologists, and are still spoken of as fish in ordinary conversation. There is one group of Mammals already noticed, the Seal family, which have some approach to the Whales both in form and habits (§ 101).

185. The general shape of the Whales is like that of fishes. The tail is, however, different in one respect. In the Whale it is flat horizontally, not vertically, as in the Fish. In swimming, therefore, it moves up and down,

while that of the fish moves laterally. Some of its motions, however, are oblique, and not wholly vertical. It is with the tail, as in the case of fishes, that the Whale mostly swims, the flippers answering the purpose chiefly of balancers. When the Whale is killed he turns over on his back, showing that it is by the action of the flippers that he keeps in his ordinary position. Though the Whale has neither hands nor feet, yet the frame-work of the flippers is much like that of a hand, as may be seen in Fig. 92, representing a flipper, and also its bones uncovered. The immense power of the tail in swimming can be judged of by its breadth, which often is 20 feet.

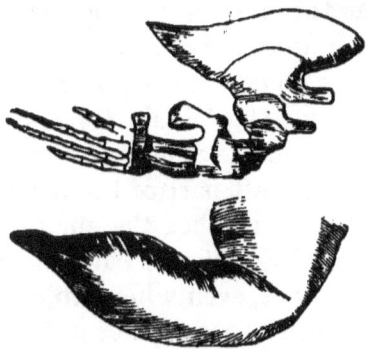

Fig. 92.—Flipper of the Whale.

186. The skin of the Cetacea is very peculiar. In other animals which have much fat, it is accumulated *beneath* the skin; but in the Whale the skin is enormously thick, and has the fat mingled with its fibres. It is this mixture of skin and fat which is called *blubber*. This is sometimes two feet thick, and weighs in some cases 30 tons; and yet, it being lighter than water, it helps to buoy up the monstrous body. When stripped of its blubber the Whale sinks at once. The mingling of the fat with the skin has two objects. One is to enable the Whale to keep its blood warm in the cold water of the frigid regions, fat being one of the best non-conductors of heat, and therefore serving to keep the heat in the body. The other is to enable the animal to bear the immense pressure of the water when it goes down to great depths.

187. Although the Whale has lungs, like terrestrial animals, it can stay under the water for a long time. It has a peculiar provision enabling it to do this. This I will explain. In the "First Book in Physiology" I showed

you that the great object of breathing is to change dark blood into red blood, and that the blood, as it returns to the heart from all parts of the body of a dark color, is sent to the lungs to be changed to red blood, before it is again distributed over the system. Red blood is necessary to every organ, to have life go on; and if it could be supplied to all the organs without breathing, then the breathing could be suspended without destroying life. Now the Whale has large reservoirs where the red blood accumulates while it is up at the surface of the water breathing. When, therefore, it goes down, every part of its body is supplied with red blood from these reservoirs. When the supply is gone, the Whale feels uncomfortable, and rises to the surface to renew the supply. The nostrils are near the highest part of the head, so that it can breathe as soon as it reaches the surface. These orifices, and also the openings of the ears, have valves, which can close so tightly that, even when subjected to the pressure of a great depth of water, not a drop can enter.

188. The nostrils are the blow-holes. The Whale has a curious apparatus for spouting. There are two large pouches under the nostrils, which can be filled with water taken in by the mouth. Here it can be retained by an arrangement of valves till the Whale wishes to spout; and then, by a forcible compression of the pouches, the water is thrown upward through the blow-holes, the valves of which are pushed open.

189. The true Whales are of two kinds or families: 1. The Spermaceti Whale, which has teeth in the lower jaw. 2. The Whalebone Whale, which has no teeth. Of the Spermaceti Whales there are two species, the most common of which, the Cachelot, or Sperm Whale, Fig. 93 (p. 111), I will notice. When full-grown it is from seventy to eighty feet long. The capture of this animal is attended with even greater danger than that of the Greenland Whale, on account of its formidable

THE WHALE TRIBE. 111

Fig. 93.—The Sperm Whale.

teeth. In the Ashmolean Museum at Oxford there is an under-jaw-bone of this whale, sixteen and a half feet in length, containing forty-eight huge teeth. It can knock a boat in pieces with its tail, or bite it in two with its jaws. In its immense head there is a very small brain, but there is a large reservoir of mingled spermaceti and oil in nearly a liquid state. A hole is cut in its head by its captors, and this mixture is baled out with buckets. By draining and boiling, the spermaceti is obtained from this separate from the oil. The blubber of this whale is thin, but yields a fine and valuable oil. The spermaceti obtained from a Sperm Whale of ordinary size amounts to about ten or twelve barrels.

190. The perfume called Ambergris is found in the intestines of the Sperm Whale. It is of the consistence of wax, is inflammable, and has a musky odor.

191. The Sperm Whales are gregarious, forming companies of some hundreds, with two of the largest as guards and leaders. Their food is fish, which they can swallow of a large size, for their throats are capacious

enough to take in a body of the size of a man. But one young is produced at a time, and this is about fourteen feet long. The milk of the mother Whale is very much like that of quadrupeds.

192. Whalebone Whales are as large as the Sperm Whales. There are two species, the Greenland Whale, and the Rorqual. The former is the best known, and is altogether the most valuable, because it furnishes the most blubber and the best whalebone. These whales have no teeth, but instead have a remarkable apparatus for taking their food, which consists of very small sea-animals of various kinds. The whalebone is the framework of the food-catching apparatus; it is in the head, in laminæ or plates to the number of three or four hundred. All of these are fringed with fibres extending down into the mouth. Now, when the Whale feeds, it rushes through the water with its huge mouth wide open, throwing out the water that enters the mouth by spouting through the blow-holes. The consequence is, that as the water passes through the fringes, the little animals in it are caught by them, and then are swallowed. The throat, in contrast with that of the Sperm Whale, is so narrow, that what an ox could easily swallow would choke this immense animal.

193. The Dolphin family of the Cetacea includes, besides the Porpoise and the Dolphin, many animals ordinarily called Whales. They all have teeth in greater number than any other Mammals, some of them even over a hundred in each jaw. The Porpoise occurs in large numbers in all the seas of Europe, and on the coasts of America. It is abundant in our bays and large rivers. Its length is from four to eight feet. It lives on herrings, mackerel, salmon, etc. It is the most common and abundant of all the Cetacea. The blubber yields a very fine oil. Its skin is tanned, and the leather is used particularly for the upper leather of boots and shoes. It is amusing to see the Porpoises rise to the surface, and

then dive down, as they chase each other in their gambols. The Dolphin is quite as sportive as the Porpoise, and much more agile. It often follows ships in numerous herds, executing its playful movements. The stories about the beautifully-changing hues of the dying Dolphin are untrue; this voracious animal is altogether unpoetical even to death. Its colors are black and white, and the only change which occurs is that the black, after a time, becomes brown, and the white gray.

194. There are some aberrant genera of the Dolphin family. One of the most remarkable we have in the Narwhal, or Sea Unicorn, as it is commonly called, Fig. 94. Its body is from thirty to forty feet long. It has

Fig. 94.—The Narwhal.

a long, straight, pointed tusk, from five to ten feet in length. It really has two tusks, but only one of them becomes long, the other not projecting sufficiently to be seen. There is much question about the use to which the animal puts this tusk. Some suppose that its chief purpose is to dig up sea-weed for food. Others suppose that the prey of the animal is transfixed by it. It is, at any rate, a very powerful weapon, and the Narwhal has been known to thrust it into the oak timbers of a ship. This animal, formidable as it is, is often taken by the Greenlander, who obtains from it oil, food, weapons, and ropes. He uses the tusk in the manufacture of spears, arrows, hooks, etc.

195. There is a family of Cetacea called the Dugong tribe, which is so aberrant that zoologists differ as to their proper place, some associating them, on account of their thick, tough skins, with the Pachydermata, and some placing them with the Cetacea. They are herbivorous, and not carnivorous like the other families of the Cetacea, living mostly on sea-weed. They have stiff mustaches, and, when their bodies are partly out of the water, they have, viewed at a distance, a somewhat human appearance, which has given rise to the "mermaid" stories. These animals are called Sea-cows, Sea-calves, etc. One species, found in the Indian Seas, especially among the islands of the Indian Archipelago, is eighteen or twenty feet in length. In Fig. 95 you have the skeleton of this

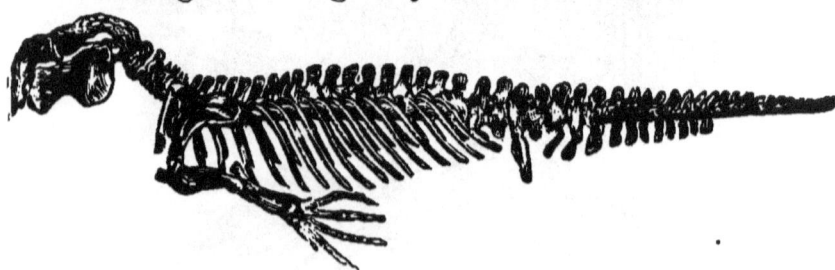

Fig. 95.—Skeleton of Dugong.

singular animal. It has, you see, no hinder extremities. The anterior extremities are paddles, like the flippers of the Whale; and the resemblance in the bones to those of the hand of man is very decided, the four fingers being present, and an attempt at a thumb. There is an animal similar to this found on the coast of Mexico and of the northern part of South America. It is, however, smaller, being but six or seven feet long, and on its paddles are short nails, by which it can drag its unwieldy body on the land to bask in the sun or to get food. All the animals of this tribe are like the Whales in their paddles, their oily skin, their horizontally flattened tail, and their fish-like shape.

Questions.—What is said of the size of animals living in water? How do the Whale tribe compare in size with terrestrial animals? How do the Cetacea differ from all other Mammals? What group of Mammals are somewhat like them? How does the tail of Whales differ from that of Fishes? What is the breadth of it? What is the chief office of the flippers? What is said of their frame-work? What is the blubber? What purposes does it serve? Explain the provision which enables the Whale to stay under water so long. What is said of the nostrils? Describe the spouting apparatus. What are the two families of Whales? Describe the Cachelot Whale. What is said of its spermaceti? How much is obtained from one Whale? What is Ambergris? What is said of the habits of Sperm Whales? What is said of the Whalebone Whales? What are included in the Dolphin family? What is said of the Porpoise? What of the Dolphin? What of the Narwhal? What of the Dugong family? What are the animals of this family commonly called? Where are they found? What is said of the structure of the species represented?

CHAPTER XII.

CHARACTERISTICS OF BIRDS.

196. BIRDS form the second grand division of warm-blooded Vertebrates. This division is separated from the first division, the Mammals, by very marked characteristics, which I will point out. 1. They are oviparous (§ 23). 2. They do not suckle their young. 3. They are covered with feathers. 4. They are constructed for flight, with some few exceptions. 5. They have no teeth, which is true of only a few species of Mammals. 6. They have bills, which is true of only one species of Mammals, the Duck-billed Platypus of Australia (§ 133). 7. They have

some peculiarities in the digestive organs, most birds having, in place of the process of mastication, a crop to soak their food and a gizzard to grind it.

197. Feathers have some resemblance to hairs, but differ from them in some important respects. A feather has commonly three distinct parts: a horny tube, or quill part; a stem proceeding from this tube; and laminæ, which are commonly joined together by barbs or teeth on their edges. The laminæ thus locked together enable the feather to press upon the air in flight. In what is called down the laminæ are very narrow, and are entirely separate.

198. The wing may be considered as a hand with a feathery appendage, so that it may press upon considerable air at once, and thus raise up the bird. Accordingly, we find that the bones of the wing are essentially the same as those in the arm and hand of man. In Fig. 96 we have the bones of a bird's wing. Comparing this with the corresponding part of the skeleton of man in Fig. 1, we have, I., the elbow-joint; II., the wrist; III., the knuckle-joint; *a*, the arm-bone; *b*, the bones of the fore-arm; *c*, the bones corresponding to those in the body of the hand; *o*, the thumb-bone; 1, 2, 3, 4, attempts at fingers. These rudimentary fingers, you see, are very different from the fingers in the wing of a Bat, Fig. 20. There they needed to be long as frame-

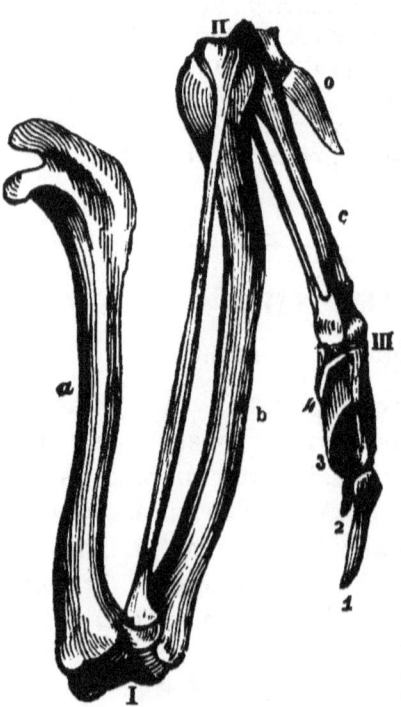

Fig. 96.—Bones of Gyrfalcon's Wing.

CHARACTERISTICS OF BIRDS. 117

work for the thin, membranous wing. But here the finger-bones are needed only for the attachment of the strong feathers, and such an extension of them as we have in the Bat would not conduce to strength, and therefore would be out of place.

199. As flying requires more strength than any thing else which the Bird does, the muscles of the wing are larger than those in any other part of the body. It is this which makes the breast so full. To accommodate these large muscles, the breast-bone has a peculiar shape. In man it is flat and small; but in the Bird it is very large, making a sort of convex buckler on the front part of the skeleton, with a ridge or keel projecting from it. In Fig. 97 you have a front view, and in Fig. 98 a side view of the breast-bone of a bird. The chief muscles

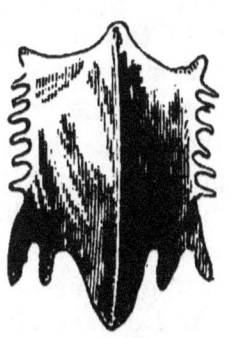
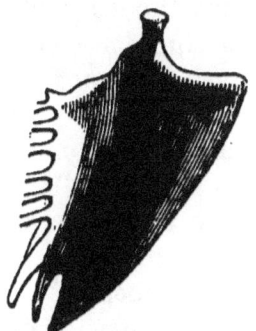

Fig. 97.   Fig. 98.

that move the wings are fastened to the keel, and spread over the breast-bone. In the birds that are cooked for the table any one can observe that this mass of muscle or meat is thickest in those birds which fly most. It is much thicker in the Pigeon, for example, than it is in the common Fowl. In those birds which do not fly at all there is little muscle on the breast, and the keel on the breast-bone is absent, as you may see in the skeleton of the Ostrich, Fig. 4. In such birds the bones and the muscles of the lower extremities are very large, while they are comparatively small in those which are much on the wing.

200. The amount of muscular power required for flight in the air is not commonly appreciated. If we look at the breadth of wing in a bird, as compared with the size of the animal when stripped of its feathers, we can have some idea of the extent of wing which a man would need to enable him to fly. And to work efficiently such enormous wings as he would require, he must have enormous muscles. Those which move the arms of the most broad-chested and brawny man are far from being large enough to enable him to fly, even if he had wings. To do this, he must have the keel on the breast-bone, like the bird, to afford an attachment for a thick mass of muscle. We see, then, why it is that all the attempts which men have made to fly have proved failures. It is not that the wings have not been properly made, but that there was not sufficient muscle to work them.

201. As flying requires such strong exertion, it is important that the Bird should be as light as possible. There is a singular contrivance for this purpose. The air taken into the lungs does not all stop there, but some of it passes thence into cells or sacs in different parts of the body, and also into many of the bones, which are hollow for this purpose. This air apparatus is in extent proportionate to the powers of flight. Thus, in the Eagle, the air goes into all the bones, while in the Ostrich and the Penguin it goes only into the thigh-bones.

202. The digestive organs of the Bird are very peculiar. They are the only animals that have a gizzard. This organ is a stomach, which has on its inside a lining as tough and hard as leather. This is for the purpose of bruising and rubbing the food, which is done by the action of very stout muscles. These constitute the bulk of the gizzard; and they are so arranged that they squeeze and rub two opposite surfaces of the inside lining against each other. The food is therefore ground in the same manner as grain is between the millstones of a flour-mill. The power of this grinding apparatus is made still more

CHARACTERISTICS OF BIRDS. 119

effectual by sand and small stones, which the Bird swallows with its food. In Fig. 99 you see the gizzard of a Turkey cut open. You observe the two semi-globular

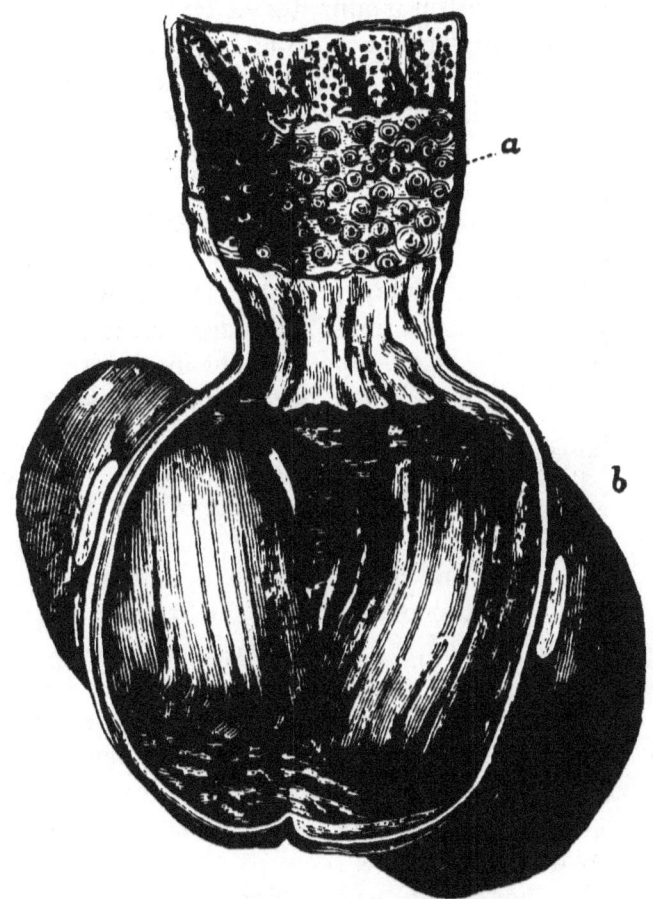

Fig. 99.—Gizzard of the Turkey.

masses of muscle, and the lining covering them on the inside of the organ. While these grind the food, the gastric juice which digests it is all the time trickling down upon it from the gullet at a, where it oozes out from a great many little openings.

203. This grinding operation of the gizzard takes the place of the mastication which is done by those animals

that have teeth; the Bird using its bill only for gathering its food, and not for masticating it. The arrangement described does not exist in full in all birds, but only in those that live on grains, termed granivorous birds. In other birds it varies according to the nature of the food. In those that live altogether on flesh, or on fishes, there is no real gizzard, but a thin and membranous stomach, for there is no need in them of any grinding and crushing process.

204. There is one part of the digestive apparatus of birds yet to be noticed. Before the food is subjected to the grinding of the gizzard, it is macerated or soaked for some time in the crop, as it is called, a sac or pouch which opens into the gullet. When the grains are first swallowed, they are passed into the crop; and when they are sufficiently macerated, they are forced out of the crop, down the gullet, into the gizzard to be ground. The crop, you see, is to the Bird what the paunch is to the Ruminant quadruped (§ 154), a convenient receptacle for the food, and a place for its maceration. In Fig. 100 you have a representation of the parts mentioned, *a* being the gullet, *b* the crop, *c* that part of the gullet where the gastric juice is made, and *d* the gizzard.

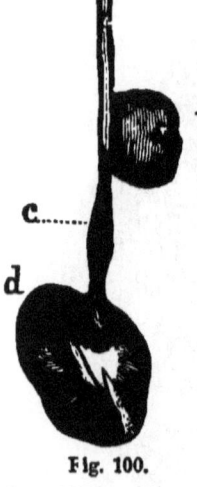
Fig. 100.

205. The incubation, or hatching of eggs, requires different periods in different species of birds. In the Humming-birds it is but twelve days, in the Canaries fifteen to eighteen, Fowls twenty-one, Ducks twenty-five, and Swans forty to forty-five. The object of sitting on the eggs is simply to provide the requisite amount of heat. The same degree provided in any other way will answer, and eggs have often been hatched by steam. The heat of the sun is sufficient to hatch the eggs of some birds

living in the tropics, as the Ostrich. The popular story about this bird is not true. There is no neglect on her part when she leaves her eggs in the sand, for when she is in a temperate climate, where the heat of the sun is not sufficient to hatch them, she sits on them. The Mound birds of Australia have a singular way of providing for heat in hatching their eggs. Instead of sitting on them, they place them in mounds of decaying vegetable matter, which they heap up for this purpose. The process of decay produces all the heat that is requisite. Most birds make nests, not to live in, but to hatch their eggs, lining them commonly with some soft material. The Eider-duck lines her nest with down which she strips from her own breast.

206. The formation of a feathered animal from the simple contents of an egg by the stimulus of heat is one of the most wonderful things in nature. When the bird is fully formed, it cuts its way out of the shell with an instrument furnished it for this purpose, a pointed scale fastened to the end of its beak. Any one can readily see this on the upper bill of the newly-born chicken. Soon after its birth this scale drops off, as the chicken has no farther use for it.

207. The senses which are most developed in birds are the sight, smell, and hearing. The sense of touch in most of them is very slight; but some, as the Duck tribe, have quite an acute sense of touch in their bills, guiding them in their search for food. The sense of taste is also, in most birds at least, very slight. The sight is generally acute, especially in birds of prey. Birds have a kind of third eyelid inside of the others, called the nictitating or winking membrane. It is very thin, and is commonly folded up in the corner of the eye out of sight, but it can be drawn over the whole front of the eye when it is needed. The bird can see through it, and the object of it is to diminish the light that enters the eye when it is very intense. It is this which

enables the Eagle, and other birds also, to look directly at the sun. The sense of smell is very acute in all birds in which it can be of service in searching for food, as, for example, in those that live on carrion. While all birds have ears, there is only one kind, the Owl tribe, that has any *external* ear. In all others there is merely an opening to the passage leading to the internal apparatus of hearing, and even this is concealed among the feathers of the head.

208. Birds are digitigrade, § 92. You can see this to be true in the case of the Ostrich, Fig. 4, if, comparing the bones of the leg with the same bones in man, Fig. 1, you begin at the thigh-bone and go downward. In Fig. 101 you have the bones of a bird's leg, *a* being the thigh-

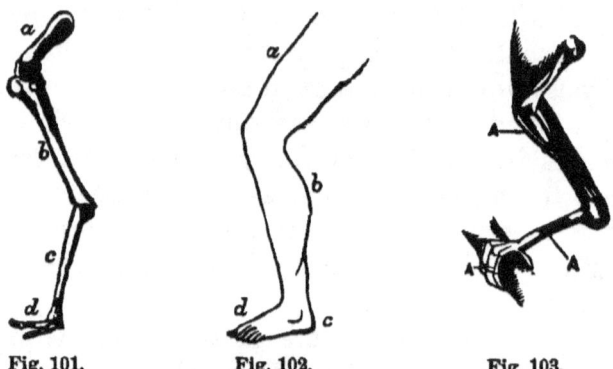

Fig. 101.   Fig. 102.   Fig. 103.

bone, *b* the bones of the leg proper, *c* the heel-bone, long and extending upward, and *d* the bones of the foot. In Fig. 102 is the outline of the leg of a man, with letters to correspond with those of Fig. 101, that you may readily make the comparison. In Fig. 103 you have the perching apparatus of birds represented, and you can see how it is that they can sleep on their perches without falling off. There is, you observe, a large muscle in front of the thigh-bone; from this a long tendon or cord, A, extends down the leg, and in the foot it divides into branches, which go to all the toes. When the muscle pulls on this the toes will all be bent, as every body knows who has played with a fowl's drum-stick, pulling upon this tendon. Now this tendon can be pulled upon in the living animal, and the toes of course be bent, without any action of the muscle. For observe, that at first the tendon is on the front of the limb, but it passes to the rear before it comes to the heel-bone. The effect of this arrangement is, that when the bird settles down in perching, the bending of the limb pulls on the tendon, and so the toes firmly grasp the perch. This arrangement is also of service to birds of prey in securing their victims; for, when they have pounced upon them, by merely settling down with all their weight, the bent claws grasp them with great force.

209. There is very great variety in the plumage of birds, the gayest colors appearing in those of tropical climates; while, on the other hand, the birds of Arctic regions exhibit none but the duller hues. The latter, however, have a much larger proportion of downy feathers to keep them warm in the midst of the severe cold. With the bright and splendid colors of the tropical birds there is no power of song, the voice being either absent or disagreeable; but in the temperate zone, while the plumage is ordinarily much less beautiful, there is great variety of song, especially in the small birds.

210. The tail is of service in flight, being moved in one way and another, so as to regulate the course of the bird. But it is not its only use to serve as a rudder; it is a part of the ornament which the Creator has given to this class of animals. Accordingly, it is varied much in its shape, arrangement, and color; and in some cases beauty seems to be aimed at rather than actual service, as in the tails of Peacocks and Birds of Paradise.

211. The instinct which leads so many kinds of birds to change their climate according to the season is a wonder and a mystery. In a temperate climate there is a multitude of birds in gardens, fields, and forests in the summer, which for the most part disappear as the cold

months come on. They migrate to the south, where the warm weather will give them the same worms or other food which they can no longer obtain at the north. Different kinds of birds have each their time for migration, which can be calculated upon with considerable accuracy. It does not depend on the degree of temperature, although it is somewhat influenced by it, as birds go north sooner when the season is early than when it is late.

212. The most mysterious part of this migration is, that the same birds will often return in the following season to the same spot which they left, although they have traversed in their migration hundreds, and perhaps even thousands of miles. This fact has been proved in the case of swallows by tying silken threads to their feet, so that there should be no mistake as to their identity. Spallanzani, a celebrated Italian physiologist of the last century, saw the same couples return to their old nests for eighteen years in succession.

Questions.—What are the two grand divisions of warm-blooded Vertebrates? In what respects do Birds differ from Mammals? What is said of feathers? What analogy does the office of a wing bear to that of a hand? Point out the resemblance between them in their frame-work. What is said of the breast-bone in birds? What is said of the muscular power required for flying? What of the inability of man to fly? What essentially contributes to the lightness of birds? Describe the digestive apparatus of birds. How is it varied in different kinds of birds? What is said of the crop? What is said of the different periods of incubation? What is the real agent of the process? What is said of the Ostrich? What of the Mound Birds? What is said of the nests of birds? What of the formation of the bird in the egg? How does it get out? What is said of the development of the senses of birds? What of the nictitating membrane? What of the ear? What is said of the legs of birds? Describe the arrangement of the perching apparatus. How is it used in taking prey? What is said of the plumage of birds? What of their powers of song? What are the two uses of the tail? What is said of the migration of birds? Is it only, or even the chief object of this, to find a suitable temperature? How much influence has the weather on migration? What is the most wonderful fact in regard to it?

CHAPTER XIII.

BIRDS OF PREY.

213. THERE are about five thousand known species of birds. They are classified mostly according to the formation of their beaks and feet, these being parts which indicate the diet and the habits. There are two grand divisions of Birds—Land Birds and Water Birds. Of the Land Birds there are five orders: 1. Raptores (*rapio*, to seize), Birds of Prey. 2. Insessores (*insido*, to sit), Perchers. 3. Scansores (*scando*, to climb), Climbers. 4. Rasores (*rado*, to scratch), Scratchers. 5. Cursores (*curro*, to run), Runners. There are two orders of Water Birds: 1. Grallatores (*grallæ*, stilts), Waders. 2. Natatores (*natator*, a swimmer), Swimmers.

214. The Raptores, or Raveners, have bills which are stout, sharp-edged, and sharp-pointed. The upper bill, or mandible (*mando*, to eat), is longer than the lower, forming a pointed hook with which the bird tears its prey. It also has a notch on each edge, which obviously can render assistance in tearing. The legs are short and stout, being very muscular, and the feet have four toes with strong claws or talons. Three of these claws are in front and one in the rear. In Fig. 104 are represented the beak and talons of a bird of prey. The strength of wing of the Raptores is adapted to their habits and their modes of taking their food. Thus the Eagle, that pounces down upon its prey, has great strength and breadth of wing; while the Owl, which approaches its prey slyly and noiselessly, has comparatively small and feeble wings.

Fig. 104.—Claw and Beak of a Bird of Prey.

215. The Raptores always live in pairs, and they choose their mates for life. It is remarkable, also, that in a large proportion of this order the females are larger than the males, probably because they have the care of the young birds, which are at first weak and blind, like the young of beasts of prey among Mammals. The colors of the plumage of this order are generally dull, brownish varied with white. They have no song, and utter only hoarse sounds. They construct their nests in a rude way in high situations, on the ledges of rocks, the tops of lofty trees, etc.

216. There are three families in this order: 1. The Falcon family, including the Falcons, Eagles, and Hawks. 2. The Vultures. 3. The Owls, which are nocturnal birds of prey (*nox*, night), the two first families being diurnal (*dies*, day).

217. The true Falcons are the most daring of all birds of prey. They are very symmetrical in form and graceful in flight. The Gyrfalcon, Fig. 105, is the most beautiful of the tribe, and the largest, it being nearly two feet long. It is found on the rocky coasts of Norway and Iceland. These birds are very courageous in defending their young. A pair of them attacked Dr. Richardson, while climbing near their nest, flying in circles around him, and now and then dashing at his face with loud screams. The Falcons were used in the once favorite sport of En-

Fig. 105.—Gyrfalcon.

gland, called hawking or falconry. Another bird of this tribe, somewhat smaller than the Gyrfalcon, was commonly used for this purpose. It is the Peregrine Falcon, found in most parts of Europe, Asia, and South America. The boldness of this bird is such that it was employed even in taking so formidable a bird as the Heron. The Falcon was held hooded on its master's hand until the Heron was aroused from its retreat; then, on being set free, it pursued the Heron aloft, each bird striving to ascend above the other. The Falcon was always victorious; and at length, with a sweep, it pounced on its victim, and both then came to the ground together. The part of the sportsman was to reach the place of conflict as soon as possible, and aid the Falcon in vanquishing its prey. So fashionable was this sport at one time in England, that persons of rank, when they appeared in public, generally had a hawking-bird on the hand.

218. The true Falcons were formerly designated as *noble* birds of prey, on account of their use in falconry, and the rest of the family were termed *ignoble* birds of prey. The Eagles are the largest birds of the latter class. There are several species, all of which have the feathers extend down on the legs even to the talons. That magnificent bird, the Golden Eagle, is among the most widely diffused of all species of Birds, being found on the Continent of Europe, in the north of England, Scotland, and Ireland, in Asia, and in North America, from the temperate to the arctic regions. It has ever been regarded as an emblem of might and courage, holding, as "king of birds," the same rank among them as the Lion does among beasts. With its powerful wings this immense bird soars to a great height, and is a grand object amid the rudeness and sublimity of the localities which it frequents. Its acute vision enables it to see its prey at a great distance, and it darts down upon it with a *swoop* or rush like that of the Falcons, but more terrific and overpowering from its greater size.

219. The nest of the Eagle is made of sticks, twigs, etc., and is generally on the ledge of some precipice, as seen in Fig. 106. In "The Land and the Book" of

Fig. 106.—Eagle and Nest.

Thomson, he describes very graphically the return of the Eagle to its nest. After making several gyrations, it poises for a moment, and then, "like a bolt, with wings collapsed, down it comes head foremost, and, sinking far below its eyrie, it rounds to in a grand parabola, and then, with one or two backward flaps of its huge pinions, like the wheels of a steam-boat reversed, it lands in safety among its clamorous children." The food of this bird consists of sea-birds, the smaller quadrupeds, as hares, rabbits, etc., and sometimes lambs, sheep, and even larger animals. In one eyrie in Germany were found the skeletons of three hundred ducks and forty hares; but the owner of the nest had undoubtedly killed, besides these, many sheep, fawns, etc., which it had stripped of their flesh, they being too large to be carried away entire to such a height.

220. The Osprey, or Fishing Hawk, Fig. 107, an aberrant species of Eagle, is spread over the whole of Europe, a part of Asia, and also portions of North America. As its name indicates, it lives on fish, which it obtains by dashing down into the water. Its nest is composed of sticks, sea-weed, grass, and turf, laid among the branches of a tree. As it is repaired and added to every year, there is sometimes enough to make a cart-load. This bird, besides living on fish, differs from the true Eagles also in having the legs covered with scales instead of feathers.

Fig. 107.—Osprey.

221. The White-headed or Bald Eagle inhabits most parts of North America. It is the figure of this Eagle which is on the national standard of this country. The food of this bird is various. While it preys on such animals as lambs, pigs, etc., it will eat fish whenever it can take it from the Fishing Hawk. If it sees this bird rise from the water with a fish in its talons, it starts off at once in the pursuit. Wilson thus describes the struggle that ensues: "Each exerts his utmost to mount above the other, displaying, in these rencounters, the most elegant and sublime aerial evolutions. The unencumbered Eagle rapidly advances, and is just on the point of reach-

ing his opponent, when with a sudden scream, probably of despair and honest execration, the latter drops his fish. The Eagle, poising himself for a moment, as if to take a more certain aim, descends like a whirlwind, snatches it in his grasp ere it reaches the water, and bears his ill-gotten booty silently away to the woods." Dr. Franklin thus speaks of this Eagle: "For my part, I wish the Bald Eagle had not been chosen as the representative of our country. He is a bird of bad moral character; he does not get his living honestly. You may have seen him perched upon some dead tree, where, too lazy to fish for himself, he watches for the labors of the Fishing Hawk; and when that diligent bird has taken a fish, and is bearing it to its nest for the support of his mate and young ones, the Bald Eagle pursues him, and takes it from him. With all this injustice, he is never in good case, but, like those among men who live by sharping and robbing, he is generally poor, and very often lousy. Besides, he is a rank coward; the little king-bird, not bigger than a sparrow, attacks him boldly and drives him out of the district. He is therefore by no means a proper emblem for the brave and honest Cincinnati of America, who have driven out all the *king-birds* from our country, though exactly fitted for that order of knights which the French call *chevaliers d'industrie.*"

222. The Secretary Bird, Fig. 108 (p. 131), derives its name from the tufts of feathers at the back of its head, having some resemblance to pens stuck behind the ear. It is allied both to the Eagles and the Falcons, but its exact place is a subject of dispute. It inhabits South Africa, Senegambia, and the Philippine Islands. It lives on snakes and reptiles, which it devours in great numbers. When attacking a snake, it uses one wing as a shield, striking the snake with the other till it is senseless; then, with a blow with its beak, it splits the snake's head, and swallows the animal. In the crop of one of those birds there were found eleven large lizards, three

Fig. 108.—The Secretary Bird.

serpents, each a yard in length, eleven small tortoises, and a great quantity of locusts and other insects.

223. The Hawks constitute a section of the Falcon family, allied to the true Falcons, but having short legs and tails. The Goshawk, Fig. 109 (p. 132), is the finest bird of this tribe, distinguished alike for its large size, its beautiful plumage, and its elegant shape. It comes nearer to the Falcons than any other of the Hawks. When it takes its prey it strikes its victim to the ground by the force with which it dashes through the air. Its food consists of hares, squirrels, pheasants, and even some quite large birds. This bird abounds all over the wooded portions of Europe, and a similar species is found in this country.

224. The Kites, another section of the Falcon family, are particularly distinguished by their long wings and forked tails. Their flight is remarkably easy and graceful, and they have the power of remaining a long time

Fig. 109.—The Goshawk.

poised almost without motion. They sweep through the air in wide circles with outspread wings, using the tail as a rudder; and they often mount up so high as to become nearly invisible. Like the Eagles, therefore, they have a wide range of vision in searching for prey; but, instead of directly rushing, like the eagles, upon their victim, they skim it, as it were, from the surface of the earth or the water, bearing it away in their talons. Their prey consists of moles, rats, mice, young poultry, and small reptiles. They will not refuse carrion. Some species perform the office of scavengers in Turkey, Egypt, India, etc., appearing for this purpose in large numbers in the streets of some of the cities. Kites performed this useful office in London as late as the times of Henry VIII. One of the most remarkable of these kinds is the American Swallow-tailed Kite, Fig. 110 (p. 133). It is found in South America, and goes as far north as 40 degrees of latitude. Its food consists of insects, small snakes, lizards, and frogs. It sweeps over the fields close to the ground, and sometimes, seizing a snake by the neck,

Fig. 110.—The Swallow-tailed Kite.

carries it off to devour it in the air, as represented in the figure.

Fig. 111.—The common Buzzard.

225. As the Hawks may be regarded as an inferior kind of Falcon, so the Buzzards may be considered as having a similar relation to the Eagles. In their flight they have neither the soar and swoop of the Falcons and Eagles, the arrow-like dash of the Hawks, nor the winding sweep of the Kites; but they sail along easily and rapidly in quest of their prey, which is much like that of the Kites and Hawks. The common Buzzard, Fig. 111,

is found in the wooded countries of Europe, and the bordering countries of Asia, and also in the fur countries of North America. There are several other species of Buzzards in this country.

226. We now come to the second great family of the Raptores—the Vultures. You have seen that the birds of the Falcon family have for their office, in the general economy of nature, to keep within bounds the number of small birds and quadrupeds, and that their head-quarters are chiefly in the cold and temperate regions. The Vulture tribe, on the other hand, have for their office to cleanse the earth from the dead bodies of animals that have died from various causes, and their head-quarters are chiefly between the tropics. Still, they are, for the most part, inhabitants of mountainous regions, some of them dwelling on the confines of perpetual snow. They descend, however, to the warm regions below in search of their food. Vultures devour bodies that Hyenas and Jackals could not reach; for none but birds can reach carcasses that are in the midst of the dense and tangled forests of the tropics, or on the steep sides of their Alpine ranges.

227. The distinguishing characteristic in the appearance of the Vultures is the absence of feathers on the head and neck, while round the bottom of the latter there is a ruff of soft feathers in a loose fold of skin, within which the bird withdraws its neck, and even the greater part of its head, when, in a semi-torpid state, as motionless as a statue, it digests the food with which it has gorged itself. This absence of feathers on the head and neck is an example of adaptation, for if they were upon this part of the body they would become exceedingly foul by contact with the carrion on which the Vulture feeds. The whole plumage of this bird is deficient in the neat and regular appearance of that of the Falcon family, and yet it can not be called a filthy animal, for it washes itself often, and spreads out its wings to the sun to be dried.

228. The Condor of the Andes, Fig. 112, is the most remarkable of the Vultures in regard to size and strength,

Fig. 112.—The Condor.

and the height to which it soars. It is about four feet long, and the expanse of its wings measures nine or ten feet; it is said to have reached in some cases even thirteen feet. Its habitual residence is ten or fifteen thousand feet above the level of the sea, and it is often seen soaring much higher than this. Besides feeding on carrion, it will often attack lambs and young goats, and when two are together, they will attack so formidable an animal as the Llama, or even the Puma.

229. The bird commonly called the Turkey Buzzard belongs to the Vulture family. It inhabits a great range of country, being found in all the warmer parts of this continent. It lives on all sorts of food. It sucks the eggs and devours the young of many species of birds, and will even eat the dead bodies of its own species. It

is daily seen in the streets of the southern cities acting the part of a scavenger. I once saw two of them, near the market in Charleston, quarreling for the possession of the entrails of an animal.

230. Some of the Vultures approach the Eagle in their form and habits. This is the case with the Bearded Vulture of the Alps, Fig. 113. It has this name from the

Fig. 113.—Bearded Vulture of the Alps.

long hair-like feathers with which each nostril is covered. As in the Eagles, the head, neck, and legs are covered with feathers, but in the characters of the eye, beak, and talons it is like the Vultures. Besides carrion, it feeds on the smaller quadrupeds which it takes as prey. It is very bold, and when very hungry will attack larger animals, and even men. It is found not only about the Alps, but also among the mountain ranges of Africa and Western Asia.

231. The Owls constitute the third family of the Raptores. They are the only birds of prey which are nocturnal in their habits, and all their peculiarities are adaptations to these habits. These I will notice. The eyes are very large, with widely opening pupils, so as to ad-

mit a great deal of light; they are also surrounded with a disk of feathers of a light color, which serves to direct the light, striking it in upon the eye. The nictitating membrane is very conspicuous, it being needed to shut out some of the light in broad day; to open its eyes widely then, and without the covering of this membrane, would dazzle the Owl exceedingly. Its head is very large and round, which is owing mostly to some cells that are connected with the organ of hearing, rendering that sense very acute; this is of essential service to it in taking its prey by night. Owls are the only birds that have an external ear, § 206. It is covered by feathers, and in some species by a sort of lid, which the bird can open or shut at pleasure.

232. The plumage of the Owls is very peculiar. It is downy, partly to keep them warm, but mostly to enable them to approach their prey noiselessly. Their flight is so noiseless that they seem borne along on the air like a tuft of down. The food of the larger species consists of hares, rabbits, fawns, birds, etc., and that of the smaller, of mice, rats, moles, small reptiles, and the larger insects. They take these either by night or in the twilight; and we find this family most abundant in those portions of the globe where the twilight is most prolonged—the cold and temperate regions. There are some aberrant species in which the habits are diurnal more than nocturnal, and, consequently, the characteristics mentioned are not fully developed in them. The *typical* species, in which the development of these peculiarities is complete, scarcely move during the day. They remain at rest upon their perch, with eyes half closed, and an amusing air of gravity; and when aroused in any way they do not fly off, but raise themselves up, and assume grotesque attitudes, making ludicrous motions.

233. The Barn Owl, Fig. 114 (p. 138), is widely diffused through the temperate regions of Europe and this country. It is a very useful animal in destroying rats

Fig. 114.—The Barn Owl.

and mice, and its presence about barns and dove-cots should be encouraged. It conceals itself in the daytime in old trees, in barn-lofts, etc., and at night sallies forth in search of its prey.

234. Of the aberrant species, the Snowy Owl, Fig. 115, is one of the largest. You see that this has not the disk around the eye which the Barn Owl has, and its shape is much like that of the Eagle tribe. This bird is found in the northern regions of both hemispheres. Its snowy

Fig. 115.—The Snowy Owl.

white plumage, from which it gets its name, makes it a very beautiful animal. But its loud voice is very terrific in the cheerless solitudes which it inhabits. It seeks its prey in the daytime, darting upon them from above, and seizing them with its stout talons, its victims being hares, various birds, and sometimes fish.

Questions.—How many known species of Birds are there? In what way are they classified? What are the two divisions of Birds? What are the orders of Land-birds? What of the Water-birds? Describe the characteristics of the Raptores. What is said of their habits? What of their plumage? Of their voice? Into what families is this order divided? What is said of the true Falcons? What of the Gyrfalcon? What of the sport of Falconry? What is the distinction between noble and ignoble birds of prey? What is said of the Golden Eagle? What of the Eagle's nest? What of its return to it? What of its food? What is said of the Osprey? What of the Bald Eagle? What does Franklin say of it? What is said of the Secretary Bird? How do the Hawks differ from the Falcons? What is said of the Goshawk? What is said of the Kites? What of the American Swallow-tailed Kite? What is said of the Buzzards? How does the office of the Vultures differ from that of the Falcon family? Where do they live? What peculiarities of appearance do they present? What is said of the Condor? What of the Turkey Buzzard? What of the Bearded Vulture? What are the peculiarities of the Owls? What are their habits? What is said of the Barn Owl? What of the Snowy Owl?

CHAPTER XIV.

PERCHING BIRDS.

235. THE second order of land-birds, that of the Insessores, or Perchers, is the most numerous and varied of all the orders. It includes all the tribes living in trees, with the exception of the rapacious birds and the climbers. Other birds perch, but the birds of this order have their feet formed especially for this. The toes are three before and one behind, the latter being on the same level with the others. They are slender and

flexible, with long and slightly-curved claws. The legs of these birds are not stout, for they are most of the time on the wing; and, on the other hand, the wings and muscles are very large in proportion to the size of the body. The plumage varies much, but, on the whole, this order excels all the others in the beauty and variety of its colors. The male is commonly larger than the female, and its colors are usually more gay. The perchers live in pairs, and build their nests in trees and bushes with some few exceptions, showing considerable skill in their construction. The singing-birds belong chiefly to this order, the only other singers being among the Scansores. As the characteristics of birds are most fully developed in this order, it is the typical order of the class.

236. There is much variety in the food of the different kinds of perching birds, and their beaks present differences corresponding to the nature of the food. Taking the form of the beak as the basis of division, there are four groups in this order: 1. *Conirostres* (*conus*, a cone, and *rostrum*, a beak), cone-billed birds, or birds having a cone-shaped beak. The greater portion of these are omnivorous, § 93, but some are exclusively granivorous. The Crows and Finches are examples of this group. In Fig.

Fig. 116.—Bill of a Grosbeak.

116 is a representation of a cone-bill, the bill of a Grosbeak. 2. *Dentirostres* (*dens*, a tooth, and *rostrum*), tooth-

billed birds. These have, as you see in Fig. 117, the head of a Shrike or Butcher-bird, a notch or tooth near the extremity of the upper bill or mandible. This is like the

Fig. 117.—Head of Shrike.

notch seen in the upper mandible of birds of prey (§ 214); and accordingly we find that those which have this notch well developed are really birds of prey, living on small birds and reptiles, as well as the insects and worms which are the common food of all this group. The Shrikes, Thrushes, and Warblers are examples of this division. 3. *Tenuirostres* (*tenuis*, thin, slender, and *rostrum*), slender-billed birds. These slender bills are specially fitted either for sucking up vegetable juices or for picking up insects. The Humming-birds are the typical birds of this group. 4. *Fissirostres* (*fissura*, a slit, and *rostrum*), gaping-billed birds. The bills or mandibles are very broad and flat toward their base, and the slit or fissure between them is carried far back under the eye. This arrangement gives them, when the mandibles are moved apart, a very broad and widely-opened mouth, as seen in the Goatsucker, Fig. 118 (p. 142). This, you see, is strongly in contrast with Fig. 116. The purpose of this conformation is to allow these birds to take insects on the wing.

Fig. 118.—Goatsucker.

which they do, passing rapidly through the air with open mouth, as in the representation of the Goatsucker. The Swallows and Kingfishers belong to this group.

I now go on to notice these groups, giving some few specimens of each.

237. The principal food of most of the *Cone-billed* Perchers consists of seeds and grains. Hence the need of the stout cone-shaped beak to pick out the seeds and to crush them. The chief families of this group are the Finches, Crows, Starlings, Birds of Paradise, Cross-bills, and Horn-bills. Most of these birds are more or less domesticable, and some of them are capable of considerable education.

238. The Finches are a very extensive family, including the Larks, Sparrows, Grosbeaks, Buntings, Linnets, etc. None of them are of large size, and some of them are very small. They have a marked general resemblance to each other in appearance and habits. They tenant fields, groves, hedgerows, and woodlands, feeding chiefly on grains and seeds, and occasionally upon insects. Many of them are great songsters. They are

commonly hardy birds, and their distribution is almost universal, some species of them being found in all parts of the globe where animals can live. They are characterized by short and thick beaks, and the two mandibles fit each other so well that, when they are together, the beak looks like a short cone, with a mere slit from the point to the base, as seen in Fig. 116. The great strength seen in the beaks of these birds is needed in opening the woody capsules covering the seeds which constitute a portion of their food.

239. I can notice but a few of this family. There are many of the *Grosbeaks* in this country, but the most beautiful and famous of them is the Cardinal Grosbeak, or Redbird. It is not only splendid in its colors, but in its song also. It is one of the prominent birds of the Middle and Southern States, and some stragglers get as far north as New England. The American *Goldfinch*, or *Yellow-bird*, one of the finest of the Finch family, lives on the seeds of hemp, the sunflower, and the thistle. From its fondness for the seeds of the latter it is often called the Thistlefinch. This bird can be educated to do many things, as drawing its drink from a glass. The *Sparrows* are an interesting group in this family. There are many species, but the two most common in this country are the Song Sparrow, one of the earliest warblers of the spring, and the Chipping-bird, so familiar to every one. There is a brown Sparrow very much like the Chipping Sparrow, but a more shy bird, brighter in color, and having a longer tail. Just before migrating in the autumn to the south, these birds, losing their shyness, come nearer to the habitations of men, and are seen flitting about in little flocks. The common *Snowbird* is one of the Finches. This hardy and numerous species, common to both continents, comes from the north in flocks into the United States in October and November, on their way south. The Bobolink, or Ricebird, as it is called as it goes south, is also one of this family.

240. The birds of the Crow family are among the largest of the Perchers. They are bold but crafty birds, showing considerable intelligence, and, when domesticated, have powers of imitation similar to those of the Parrot. They live in societies. The largest of the family is the Raven, well known in a great range of climate in both hemispheres. It has a solemn look, and has always been deemed a bird of ill omen. The Rook, so common in England, is nearly like the common Crow of this country. Of both it may be said that the good which they do in destroying grubs, which are injurious to vegetation, more than compensates for the harm which they do in pulling up the young corn or potato cuttings. The *Jays* are of the Crow family. The most beautiful of these is the Blue Jay of this country. This bird has a great antipathy to Owls, and when it discovers one, it rouses, by its boisterous vociferations, a noisy troop of birds of various kinds. The Owl receives all this with a quiet gravity, and, watching his opportunity, at length, on noiseless wing, slips away from his annoying company.

241. The birds of the Starling family are in form and habits quite like the Crow family, but are much smaller. The Meadow Lark of this country is one of them. The Baltimore Oriole, one of a numerous group in this family, is a very interesting and beautiful bird. It is called by various names: the Golden Oriole, Golden Robin, Firebird, and Fire Hangbird. Of this last name the first part was suggested by its bright orange color flashing in the light, and the latter part comes from its hanging nest, which is woven from hemp or flax. To this family belongs that singular bird of Australia, the Bower-bird, Fig. 119 (p. 145). This bird builds a bower of twigs, interwoven so as to meet above, forming a sort of tunnel. The entrance to this is decorated with any brilliant article that the bird can find, as shells and feathers. No other use has been discovered for this bower but that of a play-ground, the birds being seen to run through and

PERCHING BIRDS. 145

Fig 119.—Bower-bird.

around it in a sportive manner.

242. The bills of the Birds of Paradise are so long and slender that some naturalists have placed this family among the Tenuirostres. They are confined to New Guinea and the neighboring islands. They are distinguished for their remarkable plumes, which are of different kinds in the various species, usually consisting of

Fig. 120.—Bird of Paradise.

feathers prolonged from the shoulder-tufts or from the tail. In the species in Fig. 120 (p. 145) there is a most brilliant display of colors. The body, breast, and lower parts are of a deep rich brown; the front set close with black feathers shot with green; the throat is of a rich golden green; the head yellow; the sides of the tail have a long, full, splendid plume of downy feathers of a soft yellow color. The poetical story that this bird lives on dew, is, of course, false, and its food consists of grasshoppers and other insects, together with seeds and figs.

243. The Cross-bill family are distinguished by the crossing of the points of the beak, as seen in Fig. 121, and a horny scoop at the tip of the tongue. The bird uses these tools in obtaining the seeds of the fir and pine cones, on which it lives. The process is this: the points of the closed beak are insinuated beneath the scales of the cone, and then, by a *sidewise* motion of the mandibles, separating the points farther from each other, the scale is raised, so as to allow the horny scoop of the tongue to dislodge the seed and carry it into the mouth. It can also, with its powerful beak, extract kernels from hard shells. It will cut an apple in two to get at the pips. When confined in a cage, it very dexterously draws the ends of the wires from the wood-work, and soon sets itself free. There are three species of Cross-bills in this country.

Fig. 121.—Cross-bill.

244. The Horn-bill family are remarkable for the very large size of the beak, and for an extraordinary protuberance with which it is surmounted, as seen in the Rhinoceros Horn-bill, Fig. 122 (p. 147). This enormous bill,

Fig. 122.—Rhinoceros Horn-bill.

with its appendage, is not as heavy as it appears, for its structure is of a light, honeycomb character. The upper protuberance is hollow, and it is supposed that it serves as a sort of sounding-board, to give by its reverberations force to the roaring cry of the bird. There are several species found in India and Africa.

245. Of the division of the Perchers called Dentirostres, or Tooth-billed, there are five families: the Shrikes, or Butcher-birds, Warblers, Thrushes, Fly-catchers, and Chatterers or Waxwings. The notch in the upper mandible which makes the tooth-like projection, § 234, is not always deep, and is sometimes wanting. In such a case the proper place of the bird in the classification is known by its resemblance in other respects to the true toothbilled species. There are some, indeed, whose characteristics are so intermediate between the Conirostres and the Dentirostres, that zoologists differ as to the

group to which they properly belong. The Shrikes, or Butcher-birds, are the typical family of the Dentirostral group, having the tooth-like projection very prominent, as seen in Fig. 117. They may be styled the Raveners of the order of Perchers. In their habits they resemble the Raptorial birds. They sit motionless on their perch watching for their prey, which consists of small birds, quadrupeds, and reptiles, and the larger insects, such as grasshoppers. It is by a sudden darting movement that they take their prey. Many of them have the curious habit of impaling their victims upon thorns, showing how appropriate is their name; and they sometimes do this to so many more than they need for themselves, that some are left to dry and decay in this position. Mr. Nuttal says of the American Shrike that it has great powers of imitation, which it uses sometimes to decoy other birds into a near approach, so that it may make them its victims. Its murderous propensity is very strong. One of them, it is related by Mr. J. Brown, of Cambridge, attacked a cage in a window containing two Canaries. In its fright one of the little birds put its head through the bars, which was snapped off by the Butcher-bird, leaving the dead body in the bottom of the cage. The next day, when the cage was in the room, this bold murderer entered for another attack, but was driven off.

246. The family of Warblers consists of small birds having rather long and slender bills, with the tip slightly curved and notched. It contains a large proportion of those species which are most remarkable for their powers of song. Among them are the Bluebird and Chickadee of this country, and the Nightingale of Europe. Of this last Izaak Walton thus quaintly speaks: "But the Nightingale, another of my airy creatures, breathes such sweet, loud music out of her instrumental throat, that it might make mankind to think that miracles are not ceased. He that at midnight, when the very laborer sleeps securely, should hear, as I have very often,

the clear airs, the sweet descents, the natural rising and falling, the doubling and redoubling of her voice, might well be lifted above earth, and say, Lord, what music hast thou provided for the saints in heaven, when thou affordest bad men such music on earth!" Mr. Wood remarks of the song of this bird, "It must be borne in mind that not only in this bird, but in other singing birds, the male is the vocalist, so that Milton's address to the 'sweet *songstress*' is, unfortunately, not quite so correct as poetical; a misfortune of frequent occurrence."

247. The Warblers are spread over almost the entire globe. Audubon reckons forty-four species on the American continent. Their great office seems to be to prevent the too great increase of the insects which are found on twigs and leaves, in buds and flowers, and in the crevices of trees, their bills being well adapted for their capture. They are generally migratory birds, going south in the autumn, when the insects disappear, and coming north again in spring, when their natural prey, awakened to activity, come forth from their winter-quarters.

Questions.—What does the order of Insessores include? What are the characteristics and habits of birds of this order? Give the three groups of Perchers, and their chief characteristics. What is the food of the Cone-billed group? What families does it include? What birds does the Finch family contain? What are their characteristics and habits? What is said of the Grosbeaks? Of the Yellow-bird? Of the Sparrows? What other birds are mentioned as belonging to this family? What are the characteristics of the Crow family? What is said of the Blue Jay? What of the Starling family? What of the Baltimore Oriole? Of the Bower-bird? Of the Birds of Paradise? What is said of the Cross-bill family? What of the Horn-bill family? What are the families in the division of Dentirostres? What is said of the notched bill as a distinguishing characteristic? What birds form the typical family of this group? What are their habits? What is said of the American Shrike? How are the Warblers characterized? Mention some of them. What is said of the Nightingale? What is said of the distribution of the Warblers? What of their office?

CHAPTER XV.

PERCHING BIRDS—*continued.*

I now go on to consider the remaining families of the Dentirostral group of the Perchers.

248. The Thrushes form a very numerous and diversified family. They are almost universally distributed, appearing in nearly every variety of climate. Besides the insects on which they live in common with the Warblers, they eat, also, snails, earthworms, and various berries. They are generally larger than the Warblers. Many of them are celebrated songsters. The American Mockingbird is one of the most prominent. This bird is very abundant in the warmer portions of the United States, and is found also in the northern portions of South America; it is sometimes found as far north as Rhode Island, but not far inland. Like the Nightingale of Europe, it has a dull plumage, but it is graceful in form, and with its animated, active air while singing, has then considerable beauty. Its natural notes are very fine, and it has powers of imitation surpassing in variety every other bird. It seems to take special delight in abrupt transitions among sounds that are totally unlike, passing, for example, from the creaking of a wheelbarrow, or the sound of a saw, to the sweet song of a Canary.

249. To this family belongs the American Robin. This bird is very widely diffused in this country, being found in Oregon, and even at Nootka Sound. It is a very familiar bird, and a favorite one, partly because it comes so early in the spring, and is so late in emigrating south, and partly from its having the same name with the European Redbreast, in whose praise there has always so much been said, both in prose and poetry. The two

must not be confounded; the European bird belongs to a different family, the Warblers; it is smaller, and has greater compass and variety of song. The American Robin, however, can be educated to imitate various birds, and even to sing tunes, and it is amusing to hear it pipe out so solemn a strain as that of Old Hundred. The European bird is much more familiar than our Robin, sometimes, in winter, tapping at the window, or even entering the house in search of crumbs.

250. One of the most singular of the Thrushes is the Dipper, or Water Ousel, Fig. 123. This is found in En-

Fig. 123.—The Water Ousel.

gland, and also on the Continent of Europe, chiefly in hilly places where there are clear and rapid streams. It is a great diver, and has the habit of dipping and rising many times in succession, which gives it its name. There is an American Dipper, almost the counterpart of the European one; it is found in the western part of North America. It is very fully described by Nuttal, who says that it flits about our streams with gravelly beds, occasionally diving for its prey, and that "in the very depths of winter and in early spring it contributes to cheer its wild and dreary haunts by its simple, clear, and sweetly-warbled notes, somewhat resembling those of the young Song-thrush."

251. There are many birds of this family that it would

be interesting to notice, but I will speak of only two or three more. The Catbird, so familiar to us, is a beautiful singer, but when provoked or alarmed, utters a disagreeable mewing sound, from which it gets its name. The Brown Thrush, or Thrasher, is a songster of great sweetness and compass. So also is the European Blackbird. This must not be confounded with our common American Blackbird, which belongs to the Crow family.

252. The family of Fly-catchers is comparatively a small one in regard to the number of its species, but it is quite widely diffused. They derive their name from their skill in catching insects as they fly. For this purpose the bill is quite broad at the base, so that when the mandibles are separated, the mouth presents a wide opening. In this respect they approach the division of Fissirostres, § 236. Like them, also, they have bristles about the mouth at the sides; and their legs are small and weak, as they are mostly on the wing. The most prominent of this family in this country is the Kingbird, one of the most bold and brave of all birds. Its disposition to drive off all other birds from the neighborhood of its nest, and keep sole possession of what it considers its own domains, has given this bird its name. It will attack even such large birds as Crows, Hawks, and Eagles, mounting above them, and darting down upon their backs, and by this continual annoyance will succeed in driving them off. It will sometimes pursue one of these birds a long distance, over a mile, and then return to the neighborhood of its nest with the proud air of a conqueror, uttering rapidly its shrill and triumphant notes. I have sometimes been amused with the boldness of this bird in flying in quick darts close to my head as I approached the tree where it had built its nest. Some of the birds manage, by agility or some cunning expedient, to escape the attacks of this tyrant. "I have seen," says Wilson, "the Red-headed Woodpecker, while clinging on the rail of a fence, amuse himself with the

violence of the Kingbird, and play bo-peep with him around the rail, while the latter, highly irritated, made every attempt, as he swept from side to side, to strike him, but in vain." The Phebe-bird, which utters its *pe-wee* so continuously, is one of this family. There are eight species of Fly-catchers called Greenlets, which are familiar to this country. Their principal colors are various shades of green. One of them, from using bits of newspaper in making its nest, is sometimes called *Politician*.

253. The species of the family of Chatterers, or Waxwings, are few. The Bohemian Waxwing, Fig. 124, is diffused over Europe, and appears in England, so that its local name, accidentally given it, is not appropriate. With its silken tuft of feathers on its head, and the general silken appearance of its plumage, it is a beautiful bird, but its song is weak, as is that of all the Chatterers. There is a corresponding species pervading North and a part of South America, commonly called the Cedar-bird, or Cherry-bird. At the approach of winter the Cedar-birds leave the far north in companies of from twenty to a hundred, and go as far south as the confines of the equator. They reappear in the Northern and Eastern States in April, before the cherries and mulberries, their favorite fruits, ripen. Although they eat these fruits, they more than repay us by devouring quantities of cankerworms and other destructive insects. The Waxwings have their name from a peculiar ornament on their wings. Some of the feathers have appendages resem-

Fig. 124. - Bohemian Waxwing.

bling red sealing-wax in color. The wing is represented in Fig. 125.

Fig. 125.—Wing of Waxwing.

254. The third division of the Perchers is that of the Fissirostres. The characteristics of this tribe were mentioned in § 236. These, as you have seen, appeared to some extent in some of the Dentirostres, especially the family of Fly-catchers. The adaptation of the wide, gaping mouth, with its bristles at the sides, to the capture of insects in flight, is obvious. Some of the larger species of this tribe, however, live on fish. There are six families—the Goatsuckers, the Swallows, the Todies, the Trogons, the Kingfishers, and the Bee-eaters.

255. The Goatsuckers, of which you have an example in Fig. 118, are for the most part nocturnal, and they have the soft plumage and dull colors so characteristic of those nocturnal birds of prey, the Owls. They sally forth in the evening when the Fly-catchers and Swallows have retired to rest, and, like the Bats, skim about in the air, mostly near the ground. But while the Bats capture such hard-cased insects as beetles, the Goatsuckers take into their gaping mouths the soft-bodied moths. When these are once in the mouth they can not escape, for the bristles fence them in, and the thick saliva which is there envelops them. The foot of this bird is curiously constructed. The hind toe, as in the Owls, can be brought

forward, and the claw of the longest anterior toe has a long, comb-like projection, as seen in Fig. 126. The use

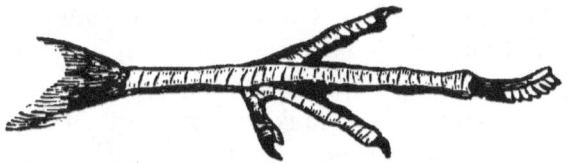

Fig. 126.—Foot of European Goatsucker.

of this is not ascertained. There is but one species of this family known in Great Britain, and this appears in all parts of Europe. There are several species in this country, one of them being the Whippoorwill and another the Night-hawk. They all have the mottled colors and large dark eyes which are so characteristic of night birds.

256. The Swallows are characterized by great power of wing, wide mouths, and short legs. The plumage of their bodies is firm and close, their wing feathers are long, stiff, and pointed, and their tails are long and forked, all which are adapted to great speed. The Swift, Fig. 127, called "Jack Screamer," is the largest of British Swallows. It spends most of the day on the wing

Fig. 127.—The Swift.

wheeling with wonderful velocity, occasionally soaring very high, and uttering its shrill screams. It captures great quantities of insects to give to its young, retaining them in a kind of pouch under the tongue. Our Chimney Swallow is one of the Swifts. It is a social bird, appearing in flocks, and making its nest in tall hollow trees or in unused chimneys. It is amusing to see them go into a chimney. The flock wheels round and round, and as they come down near the chimney those that are lowest drop in at each turn till the whole have descended. The Bank Swallow, or Sand Martin, which we see so often making holes in sand-banks with its awl-shaped bill, has its counterpart in Europe. The Martins, which so familiarly inhabit the boxes set up for them by man, are Swallows. Appearing in the extreme south of the United States the first part of February, they arrive in New England the latter part of April, and in May they are seen as far north as Hudson's Bay. They begin to emigrate from thence southward in August.*

257. The Todies are birds of gaudy plumage and rapid flight, restricted almost entirely to tropical regions.

---

* There is one species of Swallow which furnishes a singular article of diet, highly prized by the Chinese. This article is the nest of the bird. The chief material of which the nest is composed has been a subject of much dispute, some supposing it to be a kind of sea-weed, and others a substance derived from the spawn of fishes. "It is now ascertained," says Carpenter, "that this substance is secreted by enormously developed salivary glands; a few fragments of grass, hair, and other substances are generally mixed with it. The purest nests consist almost entirely of gelatinous matter, which, dissolving readily in water, is employed in making rich soups and gravies. The collecting of these nests is a proceeding of great danger; but a large number of persons are employed in it, as may be judged from the quantity sent to China. About 27,000 lbs. are annually transmitted from Java, and these are of the best quality. A still greater quantity is obtained from the Suluk Archipelago, and much, also, from Ceylon and New Guinea. It is calculated that about 30,000 tons of Chinese shipping are engaged in the traffic, and that the value of their freights is above £280,000."

PERCHING BIRDS. 157

258. The Trogons have great brilliancy of plumage, the usual tint being a golden green, contrasted boldly with scarlet, black, and brown. They are found in the tropical parts of Asia and America. The Resplendent Trogon of Mexico, Fig. 128, is the most gorgeous of all

Fig. 128.—Resplendent Trogon.

of them, its whole upper surface being of the richest metallic golden green, the breast and under parts of a bright crimson, and the tail being covered by long, soft plumes of various colors. These plumes were used by the ancient Mexican nobles as ornaments for their head-dresses.

259. The Kingfishers feed upon fish, which they take by diving. There is but one species in this country. There is a species in Europe of a similar character, but with brighter hues. It is described as being exceedingly beautiful, as, with the metallic glitter of its plumage, it glides along the bank of a river, or darts into the water. Of

the Bee-eaters there are none in this country. There is a considerable number of species in Africa, Asia, and Australia.

260. The Tenuirostres have long slender bills, intended either for collecting the honey in the nectaries of flowers, or for the capture of small insects, of which, whenever we examine flowers, we see so many in and around them. Their wings are commonly long, but the feet are slender, showing that they are to be mostly on the wing. They are, for the most part, small and of delicate form, and have great variety and brilliancy of plumage. They are almost entirely confined to the torrid zone. There are five families: the Humming-birds, Sunbirds, Honey-suckers, Hoopoes, and Creepers.

261. The Humming-birds, of which there are three hundred species, are exclusively confined to America. All but two or three are tropical birds. They are the smallest and most brilliantly colored of the feathered race. Their variety of shape may be judged of by the few spe-

Fig. 129.—Humming-birds.

cies represented in Fig. 129 (p. 158). The muscles of their wings are larger, in proportion to the size of the body, than those of any other bird. Hence their extraordinary power of flight, enabling them to dart with the velocity of an arrow, or to remain suspended in the air over a flower while they extract the honey or take the insects which are there. The humming sound, from which their name comes, is produced by the exceedingly rapid movement of the wings. The tongue is a curious instrument, being split into two tubular filaments, which can be suddenly darted out to a considerable distance. Our common Northern Humming-bird, Fig. 130, comes north as late as May. The male bird has a changeable ruby-colored throat. There is a very brilliant species found as far north on the western coast of America as Nootka Sound, the male having a crimson and copper colored throat. Nuttal speaks of it as seeming like "a breathing gem or magic carbuncle of glowing fire" as it flies about in search of its food.

Fig. 130.—Northern Humming-bird.

262. While the Humming-birds are peculiar to the New World, the Sunbirds are peculiar to the Old, almost rivaling the former in brilliancy of plumage, and resembling them in their general habits. They have similar tongues; but in gathering their food they alight, and never hover over a flower as the Humming-birds do. They differ from the Humming-birds in one respect very decidedly—they are generally agreeable songsters, while the voice of the Humming-birds is nothing but a shrill

cry. These birds range over Africa, Asia, and the Pacific Ocean. In the Hawaian Islands their feathers are highly esteemed as ornaments of head-dresses, and command a high price.

263. The Honeysuckers are an aberrant family peculiar to Australia and the neighboring islands. The Hoopoes are a still more aberrant family. The European Hoopoe, Fig. 131, with its crest, which it can raise or depress at pleasure, is a very beautiful bird. It is quite abundant in France. The Creepers are still another aberrant family, some species of which verge toward the Perchers, especially the Warbler family, and others toward the next order to be considered, the Scansores. The common Creeper of Europe,

Fig. 131.—European Hoopoe.

Fig. 132.—Creeper.

Fig. 132 (p. 160), is supposed to be of the same species with the common Creeper of this country. This pretty little bird may be seen running spirally up the trunks of trees, probing the bark here and there with its bill in search of insects that harbor in the crevices. To this family belong the Nuthatches and the Wrens, of both which there are several species in this country.

Questions.—What is said of the Thrush family? What of the Mocking-bird? What of the American Robin? What of the Water Ousel? What other birds of this family are noticed? What is said of the family of Fly-catchers? What of the Kingbird? What other birds of this family are mentioned? What is said of the family of Waxwings? What is the third division of the Perchers? What are the chief characteristics of the birds of this division? What is said of the Goatsuckers? What of their feet? What species are mentioned as belonging to this country? What is said of the Swallows? What of the Swift? Of the Chimney Swallow? Of the Bank Swallow? What is said of the Todies? What of the Trogons? What are the characteristics of the Tenuirostres? What families are in this group? What is said of the Humming-birds? What of their wings? Of their tongues? Of their humming? What is said of a species found on the western coast of America? What is said of the Sunbirds? What of the Honeysuckers? What of the Creepers?

CHAPTER XVI.

CLIMBING, SCRATCHING, AND RUNNING BIRDS.

264. We come now to the third order of Land Birds, the Scansores, or Climbers. They have four toes, two directed forward and two backward. Spending most of their time in climbing, the muscles of their lower extremities are made strong for this purpose; and, on the other hand, as they have little need of flying, the muscles of their wings are small. The order includes four families: the Parrots, Toucans, Woodpeckers, and Cuckoos. There are such marked differences between these families, that it would seem that some of them ought to be reckoned

162 NATURAL HISTORY.

as separate orders; but they all agree in their adaptation to climbing, and therefore they are classed together in the order Scansores.

265. The Parrots are characterized by their short, hard, arched beaks, and their thick fleshy tongues. They are natives of the tropical and warmer temperate regions in both hemispheres. They are remarkable for their educability and their power of imitation in the use of the voice. They have greater prehensile power than any other birds, using the beak as well as the feet in grasping. On account of this power, their intelligence, and their arboreal habits, we may consider the Parrot tribe as holding a situation among birds like that which the Monkey tribe holds among the Mammalia.

266. The Toucans, of which one species is represented in Fig. 133, are all natives of South America. Their enormous bills are made light in the same way as those of the Horn-bills (§ 244), by being of a honeycomb structure

Fig. 133.—Toucan.

The Toucan seems to be omnivorous, but is very fond of mice and small birds, which it kills by a powerful squeeze of its bill. When sleeping, it takes special care of its bill, packing it away among its feathers, so that the bird presents the appearance of a great feathery ball.

267. The Woodpeckers, so appropriately named, are widely diffused, being found in all quarters of the globe except Australia. There are eight species in this country. They live on insects and grubs, which they bore for in the bark and wood of trees. In Fig. 134 you have the attitude of the Woodpecker as he bores. The bill is

Fig. 134.—Woodpecker.

long, sharp, and stout; and with his powerful feet he holds on firmly, while he drives in his bill with all the force which his body can give to it. The sound produced by this operation is very much like that of a watchman's rattle. When an insect or grub is reached by this boring, it is drawn out by the tongue, which is specially adapted to do this. It is very long, and its sharp point is barbed with several filaments, and has upon it a gum my secretion. If the insect be of any size it is impaled, and if very small, this glutinous substance makes it adhere to the tongue.*

268. The Cuckoo family is quite an extensive one, consisting, for the most part, of inhabitants of the warmer regions. The species which in spring migrates to Great Britain, and is so common there, has the curious habit of laying its eggs in the nests of other birds of various kinds, making them perform for her the incubation needed to hatch them. The Cuckoo of this country, though very similar in most respects, has no such habit. The young European Cuckoos seem to catch the spirit of the parent, for they contrive to cast out of the nest the young of the bird by which they have been hatched. But, as they do this slyly, the foster-mother, knowing nothing of it, does not cease her tender care of the intruders.

269. The fourth order of birds is that of the Rasores, or Scratchers. The food of these birds consists chiefly of grains and seeds, and they accordingly pass most of their time on the ground. They differ in this respect from the birds that we have already noticed, which live mostly on the wing or on trees. Accordingly, the Rasores have little power of flight, and the muscles of the wings are much smaller in proportion to the size of the body than those of the Perchers and other birds of flight. Their legs are sufficiently long to enable them to walk well, and their feet are armed with short stout nails fitted for scratching in search of food. As their food is hard, and they have no teeth for masticating it, there is a crop for macerating it, and a gizzard for reducing it to pulp.

---

* It is stated by Mr. Wood that these birds do not injure trees— that the insects which they seek for are in decayed branches and stumps, and, guided by instinct, the Woodpecker bores only in these This, however, is not so, and I have this summer seen in my garden a thrifty pear-tree most curiously marked by the borings of this bird. The holes were up the trunk and out upon some of the branches in horizontal rows, from five to eight in each row.

This digestive apparatus, described in Chapter XII., is seen very completely developed in the birds of this order.

270. Most of the Scratchers do not associate in pairs. The male birds have nothing to do with taking care of the young, which are hatched with their eyes open, and are generally able to run about at once in search of food, instead of being dependent for some time on the parent for their supply. They can, for the most part, be domesticated, and they are the most useful to man of all the birds, affording him quite a large portion of his food. The plumage of the male birds is usually gay, and they often have crests or some other ornaments on the head. The females commonly differ from the males in a marked manner in these respects.

271. There is a striking analogy between these birds and the Ruminant Quadrupeds in several points. In both the food is vegetable, and in both there is a special provision for breaking it up and moistening it, so that the gastric juice may readily act upon it. The crop in the fowl answers to the paunch in the Ruminant, and the crushing by the gizzard to the grinding in rumination, each following the maceration. Then, too, these birds and the Ruminants are both, for the most part, easily domesticated, and domestication produces in both great variety in breeds. There are seven families: 1. The Pigeon family. 2. The Curassows. 3. The Pheasants. 4. The Grouse. 5. The Sheath-bills. 6. The Tinamous family. 7. The Greatfoots.

272. The family of Pigeons includes both those birds called by this name and those which are called Doves. These birds differ from those of the other families of this order in pairing; in living on trees, and, much of the time, on the wing, for which they are adapted by the large size of the wing-muscles; and in having the hinder toe on level with the others, as in the Perchers, instead of being above them, as we see it in the common fowl and in all the other families. On these accounts some have been

disposed to make the Pigeon tribe an order by themselves. The Pigeons are very remarkable for their mode of feeding their young. The crop is double, forming two pouches, one on either side of the gullet, as represented at *a* and *b* in Fig. 135. Now, while the bird is incubating, a curious change takes place in the crop, and for a special purpose. Ordinarily it is thin and smooth, as seen at *a;* but when the bird is about to have young to care for, the crop becomes thick and full of little lumps, as represented at *b*. These lumps are glands, that have now become enlarged, in order to perform their duty of pouring a milky fluid into the crop.

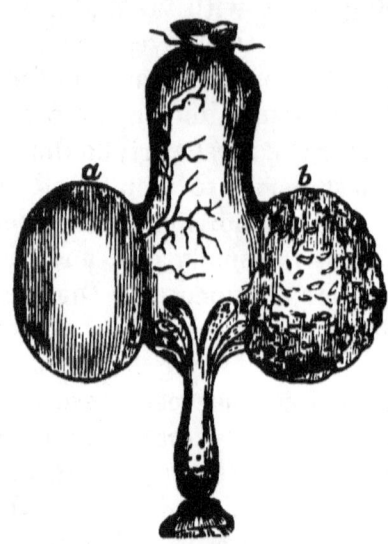

Fig. 135.—Pigeon's Craw.

The object of this is to soften the food, so that, when this is done, the bird may throw it up out of the crop and give it to its young. It is singular that the same change takes place in the male bird; and both parents, therefore, engage in feeding their offspring. The most conspicuous varieties of the domestic Pigeons are seen in Fig. 136 (p. 167). That large Pigeon, the Pouter, is able to inflate its crop with air so as almost to hide its head behind it, and it seems to be quite vain of this accomplishment.

273. This family are found in almost every part of the globe, and in some they multiply to an enormous extent. The most remarkable in this respect is the Passenger Pigeon of this country. "The associated numbers of wild Pigeons," says Nuttal, "are without any other parallel in the feathered race; they can, indeed, alone be compared to the shoals of herrings, which, descending

Fig. 136.—Domestic Pigeons.

from the arctic regions, discolor and fill the ocean to the extent of mighty kingdoms. To talk of hundreds of millions of individuals of the same species habitually associated in feeding, roosting, and breeding, without any regard to climate or season as an operating cause in these gregarious movements, would at first appear to be wholly incredible, if not borne out by most abundant testimony. The approach of the mighty feathered army with a loud rushing roar, and a stirring breeze, attended by a sudden darkness, might be mistaken for a fearful tornado. For several hours together the vast host, extending some miles in breadth, still continues to pass in flocks without diminution. At the approach of the Hawk their sublime and beautiful aerial evolutions are disturbed, like the ruffling squall extending over the placid ocean." Audubon calculated, estimating one of these flocks to be a mile in width, and allowing two pigeons to each square yard, that it numbered eleven hundred and fifteen millions, and that the quantity of food necessary for its supply must be 8,712,000 bushels per day. Their breeding-places are large forests, sometimes fifty miles long by four or five wide, in which every tree has from fifty to a hundred nests.

274. The flight of the Pigeon is very rapid. Pigeons have been killed in New York State with Carolina rice in their crops. Judging by the short time required in them for the process of digestion, it is calculated that these birds must have flown between three and four hundred miles in six hours, which is over a mile in a minute. The Carrier Pigeon has been known to fly much faster than this—nearly one hundred and fifty miles in an hour. Before the invention of the Electric Telegraph this bird was extensively employed in Europe for carrying messages. Some were trained to carry both from and to their residence. The letter was fastened under the wing or to its feet. The feet were bathed in vinegar to keep them cool, lest the bird should stop on the way to bathe. On starting, it rose high in the air, made two or three circular sweeps, and then darted off like an arrow for its place of destination.

275. The other six families are styled commonly the true *Gallinaceous* birds, from *gallus*, cock, and *gallina*, hen. The Curassows are peculiar to the tropical part of South America. Some species are as large as Turkeys, and are much prized as food. They can be easily domesticated.

276. In the Pheasant family the hind toe is placed so high that only the tip touches the ground, and there are also commonly one or more spurs. This family includes the common Fowls, Turkeys, Pheasants, Peacocks, Partridges, etc. The common Fowl is more extensively diffused than any others, and there are many varieties produced by domestication. Its native country is India, in whose jungles it is found in great numbers living on grain and seeds. The Turkeys are natives of North and Central America. The *wattles*, which are larger in these

than in any other birds of this family, are loose folds of skin well supplied with blood-vessels. These become redder and fuller when the Turkey is excited, just as the cheeks of man are reddened in blushing. The true Pheasants are allied to the Fowls. They are found wild in various parts of Asia. The most splendid species is the Argus Pheasant, Fig. 137, a native of Sumatra, Malacca, and the southeast part of Asia. The beautiful eye-spots on its plumage suggested the name of Argus, the shepherd, who, with his hundred eyes, was set by Juno to watch Io.

Fig. 137.—Argus Pheasant.

277. The Grouse family is diffused over the northern parts of America, Europe, and Asia. They differ from the Pheasants in having no naked crests or wattles, and in the absence of brilliant colors in the plumage. They vary much in size, the Partridges and Quails being birds of moderate size, while the Cock of the Wood in Europe, and the Cock of the Plains in this country, are nearly as large as the Turkey. The California Quail is a beautiful bird, having a delicate crest of a dark color, which it can erect or depress at pleasure. The Ptarmigans are an in-

teresting portion of the Grouse family. They live in the far north in America and Europe. Their legs, and even the feet, are covered with hair-like feathers. Their plumage, like the fur of the Ermine and some other quadrupeds, changes, as winter comes on, from a rich, almost tortoise-shell color, to a pure white. The trade in Ptarmigans in the north of Europe is very extensive. The captured birds are kept in a frozen state for the dealers who come for them.

278. The Sheath-bills are a comparatively small family, found chiefly in South America. Their nostrils are surrounded by a kind of sheath, and their plumage is snowy white. The Tinamous family, also a small one, is found in the same country, where they seem to occupy the same place that the Partridges and Quails do in other countries. The family of Greatfoots is peculiar to Australia and the adjacent islands. One of them is called the Brush Turkey, from its resemblance in general form to the common Turkey. It lives in the thick brushwood of Australia. This and another bird, the Mound-making Megapode (Greatfoot), are famous for making the mounds spoken of in § 205. This latter bird deposits its eggs some five or six feet deep in its mound, and then covers them up. Its mounds are very large. One of them was found to be fifteen feet high and sixty feet in circumference. They were at first supposed to be the tombs of the aborigines.

279. We come now to the order Cursores, or Runners —the Ostriches and their allies. We commonly think of birds as being, of course, capable of flight, but here we have a class of birds which are wholly terrestrial. Nearly all of them have wings, but all that their wings can do is to assist them in running. Their wings being small, the muscles which move them are small also. Accordingly, the breast-bone is entirely destitute of the projecting keel (§ 199) which it has in other birds, this being needed only for the attachment of large muscles. In the

Cursores this bone is a smooth round shield on its breast. While the muscles of the wings are small, those of the legs are very stout—their chief power is there. The plumage differs from those of birds of flight, the laminæ of the feathers not being united together by barbs (§ 197). Such a union is needed for the pressure on the air required in flying, and therefore is omitted when there is no flying to be done.

280. This group of birds is, then, an aberrant one, and, as is usually the case in groups of this character, there are but few species. One of the most prominent is the African Ostrich. This is the tallest of all birds, reaching sometimes even to eight feet. It is found in the sandy deserts of Africa and Arabia. It is probably the swiftest of all running animals. It can be domesticated, and will easily carry two men on its back. Its nest is merely a hollow made in the sand, and the hatching of the eggs is not left to the heat of the sun, but both the male and female bird engage in the incubation. The Bushmen make of these shells water-flasks, cups, and dishes. The food of the Ostrich consists of the tops of shrubby plants, seeds, and grain. It swallows, also, stones, sticks, bits of metal, leather, etc., probably guided by instinct, as these will help the grinding of the food, as the gravel does which the common Fowl swallows. There is an American Ostrich, a smaller bird, found in the southern part of South America.

281. The Emu, Fig. 138 (page 172), a native of Australia, is nearly as large as the Ostrich, but is lower on the legs, has a shorter neck, and is more thickset in body. The wings are mere rudiments, and are concealed beneath the feathers of the body. The feathers strongly resemble branching hairs, the laminæ being at a distance from each other. The Cassowary, a native of Java and the neighboring islands, is much smaller than the Ostrich. Of all the Cursores, the Apteryx of New Zealand is the one most completely destitute of wings. It has a

Fig. 138.—Emu.

very long bill, on which it sometimes rests as an old man does upon his cane placed before him. Unlike the other Cursores, it lives on insects and worms. Its habits are nocturnal, and the natives hunt it by torchlight for the sake of its skin, which is highly valued as a material for the dresses of their chiefs. It is a curious fact that in the volcanic sands of New Zealand there have been found the bones of several large birds of this order now extinct. One of them is supposed to have been fourteen feet in height.

Questions.—What are the peculiarities of the Scansores? What are their families? What is said of their differences? What is said of the Parrots? What of the Toucans? What of the Woodpeckers? Of the Cuckoos? What are the characteristics of the Rasores? What is said of the analogies between them and the Ruminant Quadrupeds? What are the families of the Rasores? What are the characteristics of the Pigeon family? What is said of their digestive apparatus? What is said of the Pouter Pigeon? What of the flocks of wild Pigeons? What of the power of flight in Pigeons? What of the Carrier Pigeon? What are the Gallinaceous birds? What is said of the Curassows? What does the Pheasant family include? What is said of the common Fowl? What of the Argus Pheasant? How do the Grouse family differ from the Pheasants? What is said of their size? What is said of the California Quail? What of the Ptarmigans? What are the peculiarities of the Sheath-bills? What is said of the Tinamous family? What of the Greatfoots? What is said of the Cursores? What of the African Ostrich? What of the Emu? What of the Apteryx?

CHAPTER XVII.

THE WADING AND SWIMMING BIRDS.

282. WE have now arrived at the Water Birds, the Grallatores and the Natatores. The Grallatores are commonly called Waders; but, as Carpenter says, they would be more appropriately named Stilt-walkers, the real meaning of the word Grallatores, for they are all remarkable for the length of their legs, while many of them can scarcely be said to be aquatic in their habits. Those which are most decidedly aquatic have their feet partially webbed. This is probably to enable them to swim in case that they should get beyond their depth. Most of the birds of this order find their food in the water, which consists of fish, mollusks, aquatic worms and insects. Their legs are, accordingly, both long and naked, so that they may wade with facility, and their necks and bills are long, that they may reach their food. They are generally slender birds, and their wings are fitted for rapid flight. Their tails are short, and they therefore stretch out their long legs behind to act as a rudder, in place of the tails of other flying birds (§ 210). They are distributed widely over the earth, and many of them make periodical migrations north and south. There are six families: Bustards, Plovers, Cranes, Herons, Snipes, and Rails.

283. The Bustards are natives of the Eastern Conti

nent and Australia. As these birds have the stout legs of an Ostrich, and are fast runners, preferring running to flying, some naturalists place them among the Cursores; but as they have wings of considerable size, and can fly readily and far, they obviously do not belong in that order. They have some alliance to the Pheasants (§ 276), for they live in part on grain, deposit their eggs in the ground without any proper nest, and do not live in pairs. The Great Bustard, Fig. 139, is the largest of all the European birds. The full-grown male is four feet long, and weighs from thirty to forty pounds. Though once common in England, it is now rarely seen there; but it is still common in Spain, Greece, in some parts of Russia, and in the wilds of Tartary.

Fig. 139.—Great Bustard.

284. The Plover family are also good runners. They belong mostly to the temperate climates of the Old World. They are found chiefly in sandy, unsheltered shores and moors. Their wings are large, and in their flight they wheel round in circles, much like the Swifts and the Pigeons. The Oyster-catcher, extensively distributed in the Old World, is also one of the Plovers of this country. It lives on Oysters and other bivalves, having a wedge-shaped bill peculiarly fitted to open them. The Lapwing, Fig. 140 (p. 175), one of the European Plovers, is a beautiful bird. It has a crest of long black feathers extending backward, and this, with the black

THE WADING AND SWIMMING BIRDS. 175

Fig. 140.—Lapwing.

and white colors of the plumage of its body, makes it a very conspicuous bird in its flight.

285. The general shape of the Crane family you see exemplified in the common Crane, Fig. 141, which is found over a large part of Europe, Asia, and Africa. In the summer it is found in the north of Europe and Asia,

Fig. 141.—Crane.

but in winter it migrates to India, to Egypt, and other parts of Africa. It flies to a great height, and even when almost out of sight its hoarse cry is audible. It feeds on frogs, snails, worms, and grain. It is about three or four feet in length. There is a singular species of Crane in

South America, called, from its loud harsh voice, the Trumpeter. It is about the size of a Fowl, and is readily domesticated. It runs rapidly, but seldom takes the wing. There are two large membranous bags connected with the windpipe at its lower part, which are supposed to give force to the voice, being used as the full bag of air is in the bagpipe.

286. The Heron family may be considered as the typical family of this order, the birds included in it being pre-eminently formed for wading. They are found on the margins of rivers, lakes, and marshes, and live on fishes, reptiles, and sometimes small Mammalia. They have, usually, long, stout, and sharp-pointed beaks, in order to capture the fish for which they watch so patiently in the attitude represented in Fig. 142. Contrary to the habits of most of the birds of this order, the Heron builds its nest in a high tree, feeding its young with fish for five or six weeks. The common Heron, Figure 142, is spread over a great part of the Old World. The plumes of this bird were formerly worn as ornaments only by the noble.

Fig. 142.—Heron.

There is an allied species in America. The Spoon-bills, notwithstanding the form of the beak, are generally ranked in the Heron family. They live by the edges of marshes, or near the sea-shore, where there are thick bushes, and their food consists of fishes, mollusks, and aquatic insects. The White Spoon-bill of the Old World, Fig.

Fig. 143.—White Spoon-bill.

143, is nearly three feet long. There is an allied species in South America called the Boat-bill.

288. The Storks are among the largest birds of this order, but they are less aquatic than the other families. They are shaped much like the Cranes, but have not the pendent plumes in the tail. They are abundant in Europe, Asia, and Africa. They winter in the latter country. They build their nests in towers, chimneys, and steeples, or in the broadly-spreading branches of a cedar or pine. In Holland, a kind of false chimney is built by the inhabitants for these birds to make their nests in. They live on rats, mice, frogs, and sometimes carrion or offal, and for this reason they are held in esteem, especially in the Eastern countries. The Adjutant of India, which is so useful in destroying vermin and offal, is one of the Stork family. So is also the sacred Ibis of Egypt, which figures so often in their hieroglyphics.

288. The distribution of the Snipe family is very general. Their food consists of insects, worms, slugs, aquatic mollusks, etc., which they obtain by thrusting their long and slender bills into mud or moist earth. Their bills are accordingly provided with nerves, so that they may know at once whenever they strike upon their prey. The flesh of these birds is held in high esteem. The Woodcock of this country has its counterpart in Europe. The Curlews, of which you have an example in Fig. 144,

Fig. 144.—Curlew.

are characterized by their curved beaks. Birds of this genus of the Snipe family are found in the northern parts of both continents. The largest of the American Curlews, appropriately called the Sickle-bill, is over two feet in length, and has a bill from seven to nine inches long.

289. In the Avocet, Fig. 145, the bill turns *upward* in its curve. This bird gathers its food by scooping it up from the mud. In searching for it, it

Fig. 145.—Avocet.

moves its bill from one side to the other like the motions of a mower, and leaves its traces in the mud. The American Avocet, Wilson says, is called by the inhabitants of Cape May, the Lawyer, from its flippant clamor. It is sometimes called Blue-stocking, from the color of its legs. The Stilt Plovers, remarkable for the great length of their legs, are included among the Avocets.

290. The Rail family are characterized by their long toes, enabling them to walk easily over soft mud or even the leaves of water-plants. Some of the tribe have for this purpose membranous margins along the sides of the toes, so that the foot may have a considerable flat surface. The Jacanas, of which a specimen is given in Fig. 146, can walk on the broad leaves of water-plants, and, as these leaves sink a little as the foot presses on them, the bird has the appearance of walking on the water. These birds are found in Asia, Africa, and America. The specimen represented in the figure is the species found in Brazil and Guiana.

Fig. 146.—Jacana.

291. Birds of the order Natatores have a peculiar provision for swimming. They are web-footed; that is, the toes are connected together by a membrane or web, as seen in Fig. 147 (p. 180), so that the feet can be used as oars or paddles. In the act of swimming, the toes are brought near together when the foot is carried forward, and they are spread out when it is carried backward. The body of the bird is boat-shaped, so as to move through the water easily. In those which are the most aquatic in their habits, the feet are placed far back, so that they may propel the body effectively; and this gives

the peculiar waddling gait which is so familiar to us in the Duck. Their plumage is dense, and it is oiled by a

Fig. 147.—Foot of Gannet.

secretion from glands, which keeps it from being penetrated by the water. Their necks are long, to enable them to reach their food. They are the only birds, Cuvier remarks, in which the neck is longer than the legs. There are five families: Ducks, Divers, Auks, Gulls, and Pelicans.

292. The Duck family have broad bills with horny laminæ at the edges, which act as a filter, allowing the water to escape, but retaining substances which are in it. While this is going on, the tongue, which is soft and well endowed with nerves, is informing the animal what is and what is not worthy of being retained, or, in other words, selecting its food. There are nerves in the bill, also, which assist in this selection. The food is various, consisting of insects, worms, mollusks, grains, etc. These birds are distributed widely over the globe, and are usually migratory. The flights of the Wild Geese in their military order are familiar to us. From the Wild Geese and Ducks come the domesticated ones. As the Goose lives more on land than the Duck, its legs are not set so far back, and it walks better. It seems to partake of the characteristics of both the Swimmers and the Waders. As it lives so much on land, its food is principally grains and grass.

293. The true Ducks may be divided into two classes, those which frequent inland shallow waters, and those

which are found in deeper waters and in the sea. The latter are well provided for in their swimming apparatus, and are good divers also. The Eider Duck, an inhabitant of the northern regions of both hemispheres, furnishes the famous eider-down. The beautiful and graceful Swans belong to the Duck family. They are inhabitants of the east of Europe and Asia. Among the singular animals of that country so fruitful in strange things, Australia, there is a Swan, the whole of whose plumage is a jetty black. The Flamingo, Figure 148, although it has the long legs of a Crane, and so is a good wader, is commonly reckoned in the Duck family, because its feet are well webbed, and its mandibles are laminated as in the case of the true Ducks. It is found in Africa, Asia, and the warmer parts of Europe. The color of the plumage is a deep brilliant scarlet, except the quill-feathers, which are black. When a flock of these birds stand in a line, as they often do, they look like a file of small soldiers. The nest of the Flamingo is a conical heap made of mud, with a hollow place in the top. When it sits on the nest its long legs hang over the sides.

Fig. 148.—Flamingo.

294. The family of Divers have short wings, and their legs are so far back on the body that they always are erect when they stand. They live on fish, which they catch by diving. They are inhabitants of the northern regions. The Grebes, a branch of this family, are not web-footed, but have their toes separate and broadly fringed along their edges. Each toe is therefore a paddle. It is supposed that this arrangement enables the bird to swim easily where there is much vegetation in the water. The quickness with which the Grebes dive is wonderful. They have been seen to dive quickly enough to avoid the shot of a gun on hearing the report, and come up at the distance of two hundred yards. They get along very poorly on land, for they are obliged to lie their whole length and then shuffle along like seals. The Crested Grebe, Figure 149, is found in Scotland and England.

Fig. 149.—Crested Grebe.

295. The Auks have, like the Divers, very short wings, and their feet are set far back. They use their wings in swimming as the whales do their flippers and as fishes do their fins, so that they may be said to fly in the water. They use their webbed feet, also, at the same time. The Great Auk, a bird three feet in length, is an inhabitant of the arctic regions. So, also, is the Puffin, another of this family. The Auks and Puffins of the northern regions are represented by the Penguins in the southern hemisphere.

The Cape Penguin, Fig. 150, is very abundant at the Cape of Good Hope and the Falkland Islands. In the water its wings are used as fins, but on the land as front legs. When it crawls, as we may say, on all-fours, it moves so quickly that it might readily be taken for a quadruped. The rookeries of the Penguins, arranged with great regularity, though occupied by vast numbers of them, have often been described by travelers. They make a singular appearance standing on the shore in dense columns in immense multitudes. The largest species of Patagonian Penguin is four feet high, and weighs forty pounds. These birds, looked at in front, appear, with their fin-like wings hanging down like arms, as so many children with white aprons on.

Fig. 150.—Cape Penguin.

296. The Gulls, in strong contrast with the family just noticed, are distinguished by great power of flight. They are found at sea at all distances, and never at any distance inland, and they are therefore said to be *oceanic* in their habits. They obtain their food at or near the surface of the water, and so are not good divers. The Stormy Petrel, Fig. 151 (p. 184), the smallest of all web-footed birds, belongs to this family. It is distributed over every part of the ocean. It is called by the sailor Mother Carey's Chicken, and is associated in his mind with the idea of a storm, because it is so much at ease even in the most violent storms, coursing over the waves in the most sportive manner. These birds are fond of accom-

Fig. 151.—Stormy Petrel.

panying ships in their course, and in doing so, fly with the greatest rapidity in every direction, now ahead and now astern. They have the faculty of standing and swimming on the surface of the water. When any greasy matter is thrown overboard, they collect about it, and facing to the windward, they manage, with their outstretched wings and their feet patting the water, to keep themselves stationary while they eat it. In calm weather, by a gentle action of the wings, they walk along on the surface of the water with the greatest ease. It was the walking of the Apostle Peter on the water that suggested the name of Petrel for these birds.

297. To the same family belongs the Albatross, so much in contrast with the Stormy Petrel in size. This gigantic bird, weighing about twenty pounds, and having a spread of wing sometimes of fourteen feet, is an inhabitant of the southern seas. With its great power of flight, it is a grand and beautiful object as it sweeps over the surface of the water in chase of the Flying-fish. This and other fish it swallows whole, being able to appropriate in this way a fish of even four or five pounds.

298. The Terns, or Sea Swallows, another branch of this family, are like the Swifts and the Swallows of the land in their long pointed wings and forked tails. Like them, also, they take their prey on the wing. Some of them live on fish, and some on insects, like the land Swallows. The common Tern, Fig. 152 (p. 185), is found in abundance on the shores of both continents. It lives on fish, which it snatches from the water as it skims over

Fig. 152.—Common Tern.

its surface with a velocity perhaps unsurpassed by any bird.

299. The last family of the Natatores, the Pelicans, are distinguished by the length of the hind toe and its union with the other toes in the web, as seen in Fig. 147. With this extent of web they are great swimmers; and yet they often perch on trees, which the length of the hind toe enables them to do. The edge of the bills is generally toothed, by which they can hold securely the fish which they take. The true Pelicans, from which the whole family is named, have a large pouch of skin hanging from the lower mandible, which serves them as the cheek-pouches do the Monkeys.

300. The Cormorant, Fig. 153 (p. 186), is one of this family. The sac is so small in the case of this bird that it can not be called a pouch. There is a powerful hook on the end of its upper mandible. It is an excellent diver, and actually gives chase to fish under water, seldom coming up without a victim. It is a very voracious animal. Waterton gives the following account of this bird's operations in the water: "First raising his body nearly perpendicular, down he plunges into the deep, and, after staying there a considerable time, he is sure to bring up a fish, which he invariably swallows head foremost. Sometimes half an hour elapses before he can manage to accommodate a large eel quietly in his stomach. You

Fig. 153.—Cormorant.

see him straining violently with repeated efforts to gulp it, and when you fancy that the slippery mouthful is successfully disposed of, all of a sudden the eel retrogrades upward from its dismal sepulchre, struggling violently to escape. The Cormorant swallows it again, and up again it comes, and shows its tail a foot or more out of its destroyer's mouth. At length, worn out with perpetual writhings and slidings, the eel is gulped down into the Cormorant's stomach, there to meet its dreaded and inevitable fate."

301. The Tropic Bird, Fig. 154, is reckoned among the

Fig. 154.—Tropic Bird.

Pelicans. This bird is noted for rapidity and endurance in flight. It has been known to be on the wing continuously for several days and nights. It sometimes takes a nap on the back of some turtle that it finds. The Frigate Pelican, or Man-of-war Bird, is another tropical bird of similar powers of flight. Its extent of wing is enormous. "Although, when stripped of its feathers," says Wood, "it is hardly longer than a Pigeon, yet no man can touch at the same time the tips of its extended wings." Under the throat is a large pouch of a deep red color, which can be distended with air at pleasure. Both this and the Tropic Bird are fond of capturing the Flying-fish.

Questions.—What is said of the Grallatores? What are their families? What are the characteristics of the Bustards? What is said of the Great Bustard? What of the Plovers? What of the Oystercatcher? What of the Lapwing? What is said of the Cranes? What of the Trumpeter? What of the Herons? Of the Spoon-bills? What are the peculiarities and habits of the Rooks? What singular birds are mentioned as belonging to this family? What is said of the Snipes? What of the Curlews? What of the Avocet? What of the Rail family? What of the Jacanas? What are the characteristics of the Natatores? What are their families? What is said of the Ducks and Geese? What of the two kinds of Ducks? What of the Swans? What of the Flamingo? What of the family of Divers? What of the Grebes? Of the Auks? Of the Penguins? Of the Gulls? Of the Stormy Petrel? Of the Albatross? Of the Terns? Of the Pelicans? Of the Cormorant? Of the Tropic Bird? Of the Frigate Pelican?

CHAPTER XVIII.

REPTILES.

302. THE cold-blooded division of the Vertebrates comprises the Reptiles and the Fishes. In the warm-blooded division the blood of each animal has a certain natural degree of heat, which is maintained quite uniform under

exposures to a wide range of temperature in the atmosphere. Thus, in man, the natural degree is 98° by Fahrenheit's thermometer, many degrees above ordinary summer's heat. This degree is maintained even in the severe cold of the arctic regions. There are various expedients for keeping in the heat made in the blood of the warm-blooded Vertebrates. Hair and fur do it in quadrupeds, feathers in birds, and blubber in whales. Man does the same by making for himself garments of materials which are good non-conductors of heat. Now in the cold-blooded division there is less heat made in the blood, and their coverings are not calculated to retain it. These animals, therefore, have a tendency to take the temperature of the air or water with which they are surrounded.

303. I will first speak of Reptiles. These are so called from the Latin word *repto*, to creep or crawl; for, although some of this class have four feet, their limbs are generally so short that a portion of the body is dragged along upon the earth.

304. The skeleton is much more varied in Reptiles than in the warm-blooded Vertebrates. In some of the Snake group all the parts of the skeleton are absent except the head, the chain of vertebræ, and the ribs, which are very numerous, amounting, in some cases, to several hundred. While in the Snake tribe the breast-bone is wanting, in the Turtle tribe it is expanded into a large under shield, the ribs also expanding above into an upper shield.

305. Reptiles can execute less rapid and less prolonged motions than warm-blooded animals. This is because the blood which circulates in the muscles and in all their organs is less stimulating. This can be seen to be true by observing how the mode of their circulation differs from that of Mammals and Birds. That this may be clear to you, I will make use of two diagrams, showing the plan of the circulation in each.

306. The first diagram, Fig. 155, giving the plan of the

REPTILES.   189

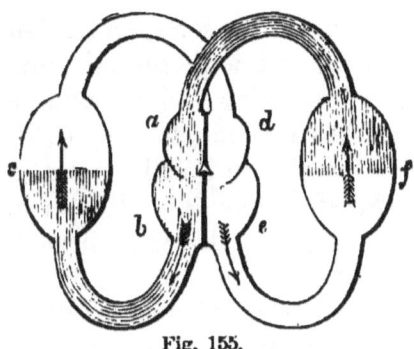
Fig. 155.

circulation in Mammals and Birds, is taken from my First Book in Physiology. In this figure, in which the shaded part shows where the blood is dark, *a* is the right auricle, which receives the dark venous blood from all parts of the body. From this it passes into *b*, the right ventricle, and this forces it out toward the lungs, *c*. Here it becomes red or arterial by exposure to the air which we breathe. It is now returned to the *left* side of the heart, and is received by the left auricle, which passes it into the left ventricle. From thence it is sent to the general system, *f*. Here it becomes dark by being used, and then returns to the right auricle, *a*, where we began to trace it.

307. The diagram, Fig. 156, is the plan of the circulation in a Reptile. In this, *a* is the right auricle, which receives the dark blood from the general system, *f*, and *d* the left auricle, which receives the arterial or red blood from the lungs, *c*. But the blood from the two auricles mixes together in one ventricle, *b*, and this mixture of red and dark blood goes alike to the lungs and to all the organs, as you see represented in the diagram. The dark shading shows where there is venous or dark blood, the light shading where there is the mixture of venous and arterial blood, and the blood is arterial or red where there is no shading.

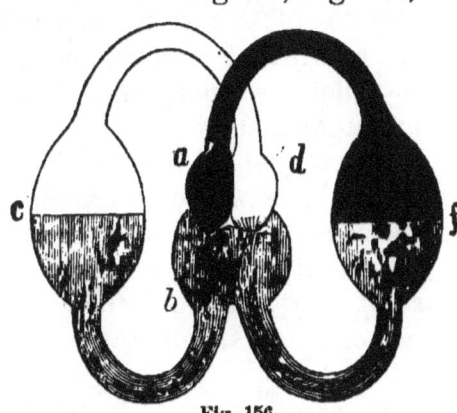
Fig. 156.

308. You see, then, that the brain, muscles, and other organs in the Reptile are stimulated with the mixture, which is not so stimulating or life-giving as pure arterial blood. It is therefore a less lively animal than those whose organs have arterial blood continually pumped into them by the heart. It therefore moves but little and slowly. Its circulation is slow, and so also is its breathing.

309. But, while life is dull in reptiles, it is not easily destroyed. They will bear being maimed to a great extent. If you destroy the brain or spinal cord of a warm-blooded animal, all signs of life soon cease; but if this be done to a reptile, motions can be excited for a long time by pricking, or other modes of stimulation. The limbs of a turtle which has been dead for several days may be made to move by pricking them, showing that there is some life in their muscles still. So, also, the two parts of a snake cut in two will move independently for some time, and the tail of a lizard will move for some hours after it is cut off. The reptiles of temperate climates crawl into some secret place as winter comes on, and go into a state of perfect torpor which lasts till spring. They are therefore called hibernating animals.

310. The brain of reptiles is very small, for they have but little thinking to do. They have no special organ of touch, and their covering is such that they can have but little sensibility in it. The sense of taste and that of smell are dull. Vision is not very acute, and the apparatus of hearing is much less complete than in the warm-blooded animals.

311. Almost all reptiles are carnivorous. The turtles and crocodiles divide their food more or less with their jaws; but the snakes or serpents swallow their food whole. In their case the throat can be so much dilated that they can swallow an animal larger than themselves.

312. Reptiles are like birds in two things: they do not suckle their young, and they produce them from eggs

They generally deposit their eggs in warm sandy places, leaving them to be hatched by the warmth of the atmosphere.

313. There are five orders of Reptiles: 1. The Turtles or Tortoises. 2. The Crocodiles. 3. The Lizards. 4. The Serpents. 5. The Amphibia.

314. The Tortoises are unlike all other animals in their covering. They are in a fortified house of bone and horn, which they carry around with them. Into this they can wholly retire when attacked. In some of the Land Turtles this covering is so jointed that they can close the openings before and behind after drawing in the head, legs, and tail, thus shutting the doors of their portable house against their enemies. The construction of this covering is worthy of examination. It is composed of two shields, an upper and a lower one. The upper one, called the carapace, has a coating of plates of horn. As the turtle grows, each plate grows by enlargement around its edge. The tortoise-shell, so much used in making combs, comes from this coating in one species. On removing this, we see that the carapace is composed of a large number of plates of bone, very nicely and firmly joined together. There is a row of eight plates through the middle, and these are appendages of the vertebræ of the back of the animal. These vertebræ you see in Fig. 6, where the lower shield is removed, so that you have a view of the under surface of the carapace. As the ribs extend from the vertebræ they expand, thus making some of its side plates. The lower shield, called the plastron, is the same thing as the breast-bone of other animals, only it is enormously large.

315. Life in these animals goes on at a low rate, and lasts a long time—in some cases even over two hundred years. Their sensibilities are dull, and it is very difficult to kill them, as they survive the severest injuries. They vary considerably in the form of their feet and of their shell, especially the former, according to their mode of

living. There are four families—Land Tortoises, Marsh Tortoises, River Tortoises, and Marine Tortoises or Turtles.

316. The Land Tortoises have short stumpy feet, somewhat like those of the Elephant, the toes not being separate, and the claws alone being apparent. They are, for the most part, inhabitants of the warmer regions, though some species live in colder climates, passing the winter, however, in a state of hibernation. Some very large species are found in and near the tropics. Thus, at the Gallipagos Islands, there are great numbers of Land Tortoises weighing over two hundred pounds. The food of the Land Tortoises is wholly vegetable. They are quiet, inoffensive animals, never making any attack, and when attacked they draw their extremities and head wholly within their portable house.

317. The Marsh Tortoises form an extensive family, diffused through the warmer countries of both continents. They are found in swamps, lakes, ponds, and small rivers. They swim easily, as their feet are expanded, and have a web between the toes. Their covering is not as firm as that of the Land Tortoises. The River Tortoises are another similar family, found in the large rivers. The American Snapping Turtle, which devours such quantities of young Alligators, belongs to this family. There is a similar species in the Nile equally destructive to the young Crocodiles. Both of these families are carnivorous, living on fish, reptiles, birds, and insects. The bony plates of the carapace of the River Tortoises are thinner than those of the Marsh Tortoises, and they are somewhat imperfect. Besides, the carapace has a coating of a leathery character in place of the horny plates of the previously noticed families. These animals are therefore sometimes called Soft Tortoises.

318. The Marine Tortoises or Turtles have their feet modified so as to be really fins or flippers. The anterior pair are most developed, as seen in Fig. 157, the Green

Turtle, and with these the animal moves rapidly through the water, they being a pair of aquatic wings. On land,

Fig. 157.—Green Turtle.

their walk is an awkward shuffle with these flippers. They are very convenient instruments, however, in scooping out holes in the sand for their eggs. Nearly two hundred eggs are laid in one nest. When laid, they are covered up with the sand. The white of these eggs, which are highly prized, does not harden in boiling. The Green Turtle, the flesh of which is considered so great a luxury, is common on the shores of most of the islands of the East and West Indies. It has been known to reach a weight of five or six hundred pounds. The tortoise-shell of commerce comes from the Hawksbill Turtle. In this animal the horny plates are large, and are arranged like shingles on a roof..

319. Of the second order of Reptiles, the Crocodiles, there are two groups—the true Crocodiles, common to both hemispheres, but most abundant in the Nile and other African rivers, and in the Ganges; and the Alligators, which are confined to America. There is not any very great difference between them; but the Crocodiles are more thoroughly aquatic than the Alligators, and

194 NATURAL HISTORY.

therefore have their hind feet more largely webbed. In the covering of both there are huge bony plates on the back and tail, rising into a prominent dentated ridge on the latter. This ridge is very elevated in the Crocodile of the Ganges, making the tail a very efficient instrument in swimming. These animals swim, in part, by the paddling operation of their hind feet, and in part by the sculling of the long, vertically flattened tail.

Fig. 158.—Crocodile.

320. There is a singular arrangement of the circulation in this order of reptiles. There are two ventricles in the heart, as in the Mammals and the Birds; but the red and dark blood are mingled together a little distance from the heart. This is not done, however, till those arteries branch off which carry the blood to the anterior part of the body. The result is, that the head and fore legs are supplied with pure arterial blood, while all the posterior parts are supplied with that mixture of red and dark blood which is supplied to *all* the organs of the other

reptiles. Why this exception is made in this order we know not.

321. In Fig. 158 (page 194) is represented the common Crocodile. Its muzzle is more elongated than that of the Alligators. That of the Crocodile of the Ganges, called the Gavial, is more prominent still, and it is terminated by a cartilaginous or gristly protuberance, in which are the openings of the nostrils. This animal is frequently twenty-five feet long, and is very formidable from its strength and ferocity. It is of great service in devouring the dead bodies of men and animals which are committed to the sacred river, and which would otherwise taint the air in their decay.

322. The Alligators or Caymans are less aquatic than the Crocodiles. They frequent swamps and marshes more than rivers. They are very dexterous in catching fish. They sometimes drive a shoal of them into a creek, and then with open mouth plunge among them. They also catch pigs, dogs, and other animals that venture too near the water.

Questions.—What are the two great classes of cold-blooded Vertebrates? How do they differ in regard to heat from the warm-blooded? What expedients are adopted in the latter to prevent the heat from escaping? What is said of the name, Reptile? What is said of the skeleton of reptiles? What of their power of motion? Describe the circulation of the warm-blooded Vertebrates. Describe that of Reptiles. What relation has the peculiarity of their circulation to their motion? What is said of their tenacity of life? What becomes of them in winter in temperate climates? What is said of the nervous system of the senses in reptiles? What is their food? What is said of the manner in which serpents eat? In what are reptiles like birds? What are the orders of reptiles? Describe the covering of Tortoises? What is said of life in them? What are their families? What is said of the Land Tortoises? Of the Marsh Tortoises? Of the River Tortoises? Of the Marine Tortoises? Of the Green Turtle? What is said of the two groups of Crocodiles? What is the peculiarity in the circulation in this order? What is said of the common Crocodile? What of the Alligator?

CHAPTER XIX.

Reptiles—continued.

323. The order of Lizards comprises a great variety of animals exhibiting some of the characteristics of the Crocodile tribe mingled with some which are peculiar to the Serpents. They resemble the former in their long body, tapering off in a tail; but, instead of the large bony plates of the Crocodiles, they have the small scales of the Serpent tribe; and, though they usually have four feet, in some of them there is but one pair, and in others the feet are so short, and so covered up by the skin, that the animal looks entirely like a snake. There is much variety in the habits of this order. Some are more or less aquatic; some are terrestrial, digging holes in the ground as places of retreat; and others are wholly arboreal. Their colors have a relation to their habits; the ground Lizards being brown and speckled, while the tree Lizards have bright colors, green predominating. When the sun wakes up the latter to activity, their quick movements make the play of their brilliant colors very beautiful. The principal families in this order are the following: 1. The Chameleons. 2. The Geckos. 3. The Iguanas. 4. The Monitors. 5. The true Lizards. 6. The Snake Lizards. 7. The Naked-eyed Lizards.

324. The Chameleons are distributed through the warmer parts of the Old World, but are not found in the New. They are distinguished from the other families by very marked peculiarities. Their bodies are flattened sideways, and there is a sharp ridge along the length of the back. Of the five toes of each foot, two are directed backward, so that the animal can grasp firmly the branches of trees in climbing. Its tapering tail is also prehensile, and is used in its arboreal mode of life as the Spider

Monkeys use theirs (§ 53). Its movements are slow. No part of it moves quickly but its tongue. This is a singular instrument. It is a long hollow tube with a swollen fleshy extremity, which is always covered with a glutinous substance. In catching insects it is darted out and returned into the mouth with a velocity which almost eludes the eye, the glutinous secretion making the insect to adhere to the tongue. The eyes of the Chameleon can be moved independently of each other, which gives the animal a strange aspect, one eye, perhaps, being directed forward, while the other is directed backward. The skin is covered with horny granulations. The changeableness of the color of the skin has been exaggerated; still, the change is perceptible through various shades from light to dark, owing to changes in the arrangement of the granules in the skin, and in the amount of blood in them. The lungs are large, and there are air-sacs connected with them in various parts of the body. When these are full of air the animal looks bloated, but the next minute it may appear lean and shrunken, having emptied these sacs. The story that the Chameleon lives on air gained currency partly from this circumstance, and partly from the almost invisible quickness of motion of the tongue, really invisible to the careless observer. The common Chameleon, Fig. 159, abounds in Northern Africa, the south of Spain, and Sicily.

Fig. 159.—Chameleon.

325. The Geckos, Fig. 160, are nocturnal Lizards, very numerous in the southern portion of Asia. They are con

Fig. 160.—Gecko.

cealed in crevices in the day, but come forth at night in search of their insect prey. They run about on the smooth walls and ceilings with perfect ease, their feet being furnished with an apparatus like a boy's sucker.

326. The Iguana family is a very extensive one. It contains over one hundred and fifty species, many of

Fig. 161.—Tuberculated Iguana.

which are among the largest of the Lizard tribe. The general aspect of the true Iguanas, which are found only in America, can be seen in Fig. 161 (p. 198), the Tuberculated Iguana. They have a long flexible tail, a crested ridge along the back and tail, and a dewlap under the throat which the animal can distend with air. The Tuberculated Iguana, sometimes reaching even six feet in length, is found in South America and the West Indies. With this family is allied the fossil Iguanodon, whose remains show that it could not have been less than forty feet in length. A very harmless little Lizard, with the terrible name of Flying Dragon, may be considered as belonging to this family, because it has the characteristic scales of the Iguanas and the dewlap. It has, like the Flying Squirrel, a wing-like expansion of the skin on each side of the body, and uses it for a similar purpose. Some of the ribs of the animal extend out as a frame-work to these wings. When running about on the branches of a tree they are folded to the side, but when it wishes to go from one tree to another, or to descend to the ground, it raises the ribs, thus expanding the so-called wings. This animal is found in the Asiatic Archipelago.

327. The remaining families of Lizards have slender tongues, which are also more or less forked. The family of Monitors includes some of the largest of the Lizards. They are graceful and agile animals, living on large insects, eggs, birds, small Mammalia, reptiles, and fish. The Monitor of the Nile, which is about six feet long, is very destructive to the eggs and the young of the Crocodile. Its name of Monitor is derived from the hissing noise which it makes when it sees a Crocodile approaching, thus giving a warning to any one that happens to be near. There are Monitors, also, in this country, which give a similar warning of the approach of the Alligator.

328. The true Lizards are bright-eyed, slender, and lively little animals, with brilliant colors, especially those

that live in verdant places. They are found in all warm countries except Australia and the Polynesian Islands. Some are natives, also, of temperate climates, passing the winter in a torpid state. The common Lizard, Fig. 162,

Fig. 162.—Common Lizard.

is only about six inches long. In all the animals of this family the tail is exceedingly brittle, snapping off like glass even with a slight touch. It grows out again, however, and if the tail be cracked without being broken off, a new tail will spring from the crack, so that the animal will have thus a forked tail.

329. In the family of Snake Lizards we find a series of forms, in which we see a gradual transition from the order of Lizards to that of Serpents. In some of these animals there are four feet, as seen in Fig. 163, the Snake Lizard of the South of Africa. Others have but two feet.

Fig. 163.—Snake Lizard.

Others still have nothing but the mere rudiments of feet concealed in the skin. Of this latter kind is the Blindworm or Slow-worm. This animal, which is about a foot in length, is as brittle as the tail of the true Lizards. The

Glass Snake of this country is also one of the same kind of Snake Lizards.

330. In the family of Naked-eyed Lizards the approach to the Serpents is still greater. Not only is the body snake-like, but the eyes are, as in the Snakes, destitute of eyelids, and covered only with a transparent portion of the skin. Most of the species of this family are natives of Australia, and only one is found in America.

331. We now come to the order of Snakes or Serpents. The grand peculiarities of this order are the total absence of limbs, the great flexibility of the chain of vertebræ, which runs through the whole length of the animal, and the covering of scales. Over the scales spreads very closely a thin delicate skin, which is shed every year or oftener, a new one forming in its place. The separation is begun at the head, and the skin, in being cast off, is turned inside out, as we sometimes turn the finger of a glove. The Serpents of temperate climates hibernate, and on waking up in the spring cast off their skins.

332. The skeleton of a serpent is very simple, consisting only of the skull, the column of vertebræ, and the ribs. There is no breast-bone. Each vertebra is united by a ball and socket joint with the one next to it. It is this arrangement that enables the animal to execute its free and graceful movements. The vertebræ, in some cases, number as high as three hundred.

333. The ordinary forward movement of the Serpent is made by the ribs, the scale which is at the end of each one of them acting as a foot on the ground. These scales being successively pushed backward against the earth, the animal is moved forward. But sometimes it gathers itself up into a coil, and then, by the sudden straightening out of its whole body, it can at once reach more than its whole length, leaping upon its prey.

334. The senses of the Serpents are not highly developed. Sight is the most perfect of all its senses. The eyes, however, are small, without eyelids, being covered

with a transparent membrane which is shed with the skin. The tongue is soft, and forked at its end, and it is not very sensitive. The smell and the touch are both rather dull. Serpents have teeth, but not for mastication. They only serve to retain their food.

335. The species of this order may be grouped in two classes—the Viperine and the Colubrine Snakes. The Viperine Serpents have a peculiar venomous apparatus. There are two teeth or fangs in the upper jaw, connected with the gland in which the poison is made. They are movable, and when the animal does not wish to use them, they lie backward, concealed along the roof of the mouth. When the serpent bites, he throws these fangs forward, and, at the same time, a muscle, pressing on the gland, forces out the venom, which passes along a canal in the fang. Most of the Colubrine Snakes are not venomous, and those which are have stationary instead of movable fangs.

336. There are two families of the Viperine Snakes— the Viperidæ and the Crotalidæ. The Viperidæ belong exclusively to the eastern hemisphere. Those of the tropical regions are the most venomous. To this family belongs the Horned Viper, so called from a small pointed horn above each eye. This is supposed to be the Asp, from whose bite Cleopatra died. The Puff Adders of Africa also belong to this family.

337. Of the family Crotalidæ, the true Rattlesnakes, Fig. 164 (p. 203), are confined to this country, but there are other species found in Asia. The rattle consists of a number of thin, horny appendages, which are loosely jointed together, and which make a rustling noise when the snake moves. The number of joints is increased, up to a certain amount at least, with each casting of the skin.

338. The Colubrine Snakes have two families—the Colubridæ or Colubers, and the Boidæ or Boas. The family of Colubers contains more than half the whole number of species of Snakes. Of the comparatively few of these

Fig. 164.—Rattlesnake.

which are venomous, the Cobra di Capello of India, Fig. 165, is the most noted. This belongs to the Hooded Snakes, so called from a peculiar arrangement of the skin

Fig. 165.—Cobra di Capello.

of the neck, by which, when the animal is irritated, it is made to take the form of a hood. While the Colubers are very widely distributed in the earth, the Boas are confined to hot climates. The latter are Serpents of enormous size and great muscular strength; and from their power of coiling round their victims and compressing them, they are able to overcome animals of very large dimensions. After destroying the life of their victims by compression, they proceed to swallow them whole; and such is the power of distention in their throats, that they can do this with men and even with cattle. The usual length of Boas is from fifteen to thirty feet, but there is a well-authenticated account of the killing of one which measured sixty-two feet.

339. There remains to be considered another order of reptiles — the Amphibia (ἀμφίβιος, *amphibios*, having a double life). They are sometimes, also, called Batrachia. These reptiles, including Frogs, Toads, Salamanders, etc., are intermediate between the other orders of reptiles and fishes. When first born, they are, like fishes, possessed of gills, and live wholly in the water. Then a series of changes takes place, the animal being at length endowed with lungs in place of gills, and fitted to live on land. This may be exemplified by reference to the Frog, which is at first a Tadpole, living in the water, having fringed gills and a long tail, with which it swims with considerable agility. It goes through a succession of changes, in which it loses its tail and its gills, and gains four legs and a pair of lungs. You will find these changes represented in my "Human Physiology," page 113. Some of the animals of this order do not lose their gills in the transformation, but, in their perfect state, have both gills and lungs. These, in the strict sense of the term, are amphibious, or double lived.

340. In their perfect or mature state the Amphibia are, in most respects, like the reptiles which we have already noticed, and therefore are properly classed with

them. They differ from them in some respects, a few of which I will notice. Prominent among these is the series of changes in passing to their mature state, of which I have just spoken. The reptiles of the other orders are covered with plates, or shields, or scales; but the Amphibia have a smooth skin, with the exception of a few species, whose scales are much like those of a fish. This skin is in many cases moist, and in some the secretion which makes it so is irritating to one who handles the animal. The Amphibia have no ribs, and therefore, not having the means of dilating the chest, must swallow air as they swallow food, directing the one to the lungs and the other to the stomach. You can therefore suffocate a Frog or any animal of this order by wedging its mouth wide open; that is, you prevent the air from going into its lungs as effectually as it is done with most other animals by closing the passage to the lungs. There is one other order of reptiles of which the same is true—the Tortoises. This is partly because the ribs are joined to the carapace (§ 314), and therefore are not movable, and partly because the plastron below does not permit that protrusion of the abdomen which we see always produced by the action of a diaphragm. No reptiles have a diaphragm, but all except the Tortoises and the Amphibia can dilate the lungs by means of their ribs. The feet of the Amphibia are without claws. Their eggs have no hard covering or shell. They are usually deposited in the water, even in the case of those that live mostly on the land. They are enveloped in a glutinous matter, which unites them in masses, or in chains, the latter looking like necklaces of black beads.

341. The tongues of the Batrachians are commonly large and fleshy. In the Frogs and Toads there is a very peculiar arrangement. The tongue is fastened to the front of the jaw, and its tip extends backward toward the throat. It is covered with a slimy substance, as the end of the Chameleon's tongue is (§ 324), and for a sim-

ilar purpose. Like the tongue of that animal, it is darted out and returned with such velocity, in catching insects, that we must be very quick of sight to see the thing done. Even the nimble Fly that comes near to the lazy Toad is not quick enough to escape its tongue.

342. The chief families in this order are, 1. The Ranidæ, or Frogs. 2. The Bufonidæ, or Toads. 3. The Salamandridæ, or Salamanders. 4. The Sirenidæ, or Sirens; and, 5. The Apoda, or Footless Amphibia.

343. The Frogs, although good swimmers, and found in the neighborhood of water, pass most of their time on land, catching insects with their tongues. They have teeth in the upper jaw. Their hind legs are long, and they are therefore good at leaping. The noisy Bull-frog is found only in North America. It lives on fish and snakes as well as insects. The Edible or Green Frog abounds in Europe, and is thought much of as an article for the table. In some places it is fattened in "froggeries" for this purpose. The Tree Frogs are arboreal, as their name indicates. To enable them to retain their position easily as they leap about among the branches, their toes have little suction pads, similar to those of the Geckos (§ 325), which, to make them the more efficient, are always covered with a glutinous secretion. Like the common Frogs, they breed in the water, and bury themselves in the mud for their winter's sleep.

344. The Toads have no teeth in the upper jaw as the Frogs do. They have also shorter legs, and therefore have less power of leaping. The skin has wart-like projections, from which an acrid fluid is secreted. The Surinam Toad, which is put by some into a separate family, is a very singular animal. It has no tongue, and its hind feet are webbed. It is found in dark corners about houses in Guiana and Surinam. Its eggs are hatched in a curious way. The male places them in little pits on the back of the female, each pit having a lid; and, when hatched, the little Toads, lifting the lids, hop out. There

is a small Frog in Venezuela that has a similar contrivance, hatching its eggs in a pouch on its back.

345. The common Newt, Fig. 166, is a specimen of the

Fig. 166.—Common Newt.

Salamander family. It feeds chiefly on Tadpoles and worms, which it eats with a peculiar quick snap. These animals are, you see, much like the Lizards in shape; but they are considered as belonging to this order, because they go through with the changes spoken of in § 339. The true Salamander is a land animal of the same general character with the Water Newt, but having a rounded tail. The stories about its being capable of living in the midst of fire are wholly unfounded.

346. The Sirens have only the anterior legs developed, and that only to a small extent. They are found principally in the marshy rice-fields of the Southern States of this country. One species sometimes reaches a length of three feet. The Footless family contains but a single genus—the Blind Newt, or Naked Serpent. Cuvier placed it among the Serpents on account of its snake-like form. But it has no scales, and it is found to undergo the metamorphosis, or change of form, common to all the Amphibia.

Questions.—What is said of the Lizard tribe? What are the chief families? Where are the Chameleons found? What are their characteristics? What is said of their tongues? Of their eyes? Of their changing color? Of their air-sacs? What is said of the Geckos? What of the Iguana family? Of the Iguanodon? Of the Flying Dragon? Of the Monitors? What is said of the true Lizards? What of the Snake Lizards? Of the Naked-eyed Lizards? What are the peculiarities of the Snakes or Serpents? What is said of their skeleton? How does a serpent execute its movements? What is said of the senses of Serpents? What are the two classes of Serpents? What is the great peculiarity of the Viperine class? What are its two families? What animals are mentioned as belonging to the Viperidæ? What is said of the Crotalida? What are the two families of the Colubrine Snakes? What is said of the Colubers? What of the Boas? What does the order of Amphibia include? From what does their name come? What is said of their metamorphosis? How do they differ from other reptiles? What is said of their tongues? What are their families? What is said of the Frogs? Of the Toads? Of the common Newt? Of the Salamanders? Of the Sirens?

CHAPTER XX.

FISHES.

347. THE Fishes constitute the second division of cold blooded Vertebrates. They are the only vertebrated animals that are fitted to live entirely in the water. All the peculiarities of their structure have reference to this mode of life. These I will proceed to point out.

348. All animals must breathe in order to live; that is, they can not live unless they have the blood exposed to the action of the air. This is as true of Fishes as it is of other animals. They breathe the air mingled with water, and can not live in water that has no air in it. This can be proved by experiment. If a fish be put into a close vessel, it soon uses up all the air in the water; and it dies if more air be not introduced into the water by unclosing the vessel. A fish dying in this way may be truly said to be *drowned*.

349. The Fish has not lungs, for these are organs which are fitted to introduce air alone to the blood. But it has *gills*, which are fitted to have the blood in them acted upon by the air that is mingled with water. These gills are fringes which are made up of very minute blood-vessels. There are commonly four of them, fixed to some arches of bone; and they are covered on the outside by a lid, called the operculum. In order to have the air in the water act on the blood in them, the water is taken into the mouth, and then passes out through these fringes. If you watch fishes in an aquarium, you will see the mouth constantly opened to take in the water, and the operculum as constantly raised to let it out.

350. When a fish is taken out of the water it really dies for want of air, although it is in the midst of a plenty of it. The explanation is this: the fringes of the gills are kept apart by the water while the fish is in its native element; but, when taken out of it, the fringes fall together, and soon become dry. When they are in this condition the blood will not circulate freely in them, and what blood is there is not acted upon by the air. In agreement with this explanation we find that those fishes which live the shortest time out of the water have their gills most exposed, while those that live a longer time have their gill-openings narrow, thus tending to keep the fringes moist. In some there is an especial arrangement for moistening them, and in such a case the fish can live in air quite a long time. Dr. Carpenter states that some fishes having this arrangement are accustomed to leave the water and crawl about in the grass or on the ground.

351. The plan of the circulation in the Fish is peculiar. In the Mammals and Birds there is a double heart, as illustrated in Fig. 155. In Reptiles the heart is double only so far as the auricles are concerned, as illustrated in Fig. 156. In Fishes the heart is single, having but one auricle and one ventricle. The blood passes from the ventricle to the gills. Here it becomes arterial blood, as

the blood of Mammals, Birds, and Reptiles does in the lungs. But, while their blood returns to the heart from the lungs before it is distributed over the system, the blood of the Fish is distributed directly from the gills.

352. The circulation of Fishes is not as active as that of Mammals and Birds, and their blood is cold like that of Reptiles. We can readily see why it is best that it should be cold. There are only two ways in which it could be *kept* warm, like that of warm-blooded animals. One is by having a covering of feathers or of fur, as in the case of animals living in air. But such a covering would interfere very much with swimming. Another way to retain the heat would be to have, like the whales, a thick layer of fat under the skin. This would be very burdensome; and, besides, man does not need such a supply of fat and oil as this arrangement in the Fishes would give him.

353. The shape of the Fish is such as to let it move easily through the water. It has, commonly, a long, spindle-like shape, with an even surface. It has no neck, chiefly because any irregularity in its surface would hinder its rapidity of motion. Its outer covering favors its gliding through the water, for it is generally composed of smooth scales, one overlapping another, like shingles or tiles. Then there is a slimy, oily secretion over the whole surface, helping it to move smoothly through the water.

354. The Fish is nearly of the same specific gravity with water. It is therefore obliged to make very little effort in going upward. It is in strong contrast with the Birds in this respect. A bird, in mounting upward, exerts great force with its broad wings and its large muscles, because it is in an element which is so much lighter than itself. But as the fish is in an element only a little lighter than itself, it needs but a small apparatus to move in it, and, accordingly, its tail and fins are much smaller in proportion to its bulk than are the wings of birds in pro-

portion to theirs. For the same reason, while man can swim with his hands and feet, he can not fly. The Flying Fish, Fig. 167, is enabled to fly by having fins which approach in extent the wings of a bird.

Fig. 167.—Flying Fish.

355. Besides, most fishes have a peculiar contrivance for enabling them to rise and fall in the water easily. It is a bladder of air which the fish has the power of compressing or enlarging at pleasure. If the fish wishes to go down rapidly in the water, it compresses this air-bladder, and so increases its specific gravity. If, now, it wishes to rise, it takes off the pressure from the air-bladder, which therefore enlarges to its former dimensions, and lessens the specific gravity of the fish, or, in other words, makes its bulk greater, while the weight remains the same. Sometimes the fish loses its power of compressing the air-bladder, and then it is so light all the time that it has no power to go down in the water. A gentleman had a Goldfish which swam with its belly upward, probably from a wrong position of the air-bladder.

356. The chief agent in swimming in the Fish is the tail, which acts like a sculling-oar, moving to the one side and the other alternately. It is terminated, for this pur

pose, with a considerable finny expansion, consisting of a skin, over a frame-work which is sometimes bony, and sometimes cartilaginous or gristly. It is constructed, therefore, very much like the wing of a bat. The fins are similarly constructed. These generally act chiefly as balancers and directors of the movement, while the sculling tail propels. That the side or pectoral fins, however, have considerable agency in propelling, can be seen very readily, if you watch the movements of fishes in an aquarium. They obviously narrow and widen as they are moved, widening when they make a propelling stroke.

357. The skeletons of Fishes are not as firm as those of other Vertebrates. In some, even, they are not real bone, but are cartilaginous or gristly. The reason of this difference is plain. As the Fish moves in an element of nearly the same specific gravity with itself, it puts forth but little strength in its movements. The points of support, therefore, for the muscles need not to be so firm as they are in animals living in air and exerting motions that require considerable force, such as springing from the ground, grasping, flying, etc.

358. We see a marked adaptation in the Fish to its mode of life in the organs of sense and the brain. Its life is passed mostly in obtaining its food and in escaping from its enemies. Its life is a lazy one compared with that of animals that run, and dig, and scratch, and climb, and fly. It shows, neither, any remarkable instincts. It therefore does not need much of a brain, for its range of thought is very limited; neither does it require acuteness in the senses to meet its wants. Its brain is therefore small, and the organs of sense are not as fully developed as in some other animals. It has little sense of touch, and it is mostly confined to the lips. The filaments which some have about the mouth are probably organs of touch, informing of the contact of bodies just as the whiskers of a cat do. The eyes of a fish are large and nearly immovable. As they are lubricated by the

water, they need no eyelids and no tear-apparatus, and accordingly have none. For protection of the eye the skin is continued over it, but it is so thin that the light is readily transmitted through it. The organs of the sense of smell are better developed than those of any other sense, and its smell is therefore acute, undoubtedly to aid it in the search of its food. Its sense of taste, on the other hand, is dull, probably because, for the most part, the food is swallowed whole, and is not detained long in the mouth.

359. Fishes are very voracious, and their food is mostly animal, few, comparatively, feeding upon vegetables. Some live on the soft-bodied animals floating in great numbers in the sea, of which I shall treat in another part of this book. Others live on shell-fish, and animals that are covered with a hard crust, such as lobsters, crabs, etc. Many fishes in fresh water live on worms and the grubs of insects. Then, too, fishes feed to a large extent on fishes that are smaller than themselves. In this respect such fishes have a resemblance to the carnivorous Mammals, Birds, and Reptiles. It is stated that, "at a lecture delivered before the Zoological Society of Dublin, Dr. Houston exhibited, as ' a fair sample of a fish's breakfast,' a Frog-fish two and a half feet long, in the stomach of which was a Codfish two feet in length. The Cod's stomach contained the bodies of two Whitings of ordinary size; and the Whitings, in their turn, held the half-digested remains of many smaller fishes too much broken up to be identified."

360. The mouth of each fish is adapted to its mode of gaining a livelihood. Some species have no teeth, but in most fishes there are several rows of them. They are commonly not confined to the jaws, but are also on the tongue, the palate, etc. Most have teeth merely for holding the food and passing it into the throat, while in others there are teeth for cutting or tearing; and in such as live on shell-fish there are teeth for crushing.

361. Fishes surpass any other class of Vertebrates both in actual number and in the number of genera and species. Seven tenths of the earth's surface is covered with water, leaving out of view the lakes and rivers. Now in the seas and oceans occupying this immense space Fishes are found, some in shallow waters, some out at sea near the surface, and others at various depths, some even to the depth of several hundred feet. There are different ranges of depth to which different species are fitted. It is among those whose range does not extend much below the surface that we have a display of bright and various colors, equaling, often, that of the Bird and Insect tribes; while those that frequent the depths are of a dull color. This difference is owing mostly to the influence of light. At what depth there is a total absence of light, and therefore, probably, of life, has not been ascertained.

Questions.—How do the Fishes differ from all other vertebrated animals? What is said of the necessity for breathing to a fish? How may a fish be drowned? What is the difference between lungs and gills? How are gills constructed and arranged? Why does a fish die when out of water? Why can some live a considerable time in air? What is the plan of the circulation of fishes? Why is it best that they should be cold-blooded? What is said of the shape of a fish? Why has it no neck? What is said of its covering? How is the motion of fishes contrasted with that of birds? What is said of the Flying Fish? How do fishes use their air-bladder? What is said of the use of the tail and of the fins in swimming? What is said of the skeletons of fishes? How is the Fish adapted in its nervous system and its senses to its mode of life? What organs of touch do we see in some fishes? What is said of the organs of vision in fishes? What of their sense of smell? What is said of their food? Give the statement of Dr. Houston. What is said of the mouths of fishes? How do fishes compare with other vertebrates in number and variety? What is said of the various depths in which they live? What influence has this on their color?

CHAPTER XXI.

FISHES — *continued.*

362. SOME fishes are fitted to live in fresh water, and some in salt, while others can live in both equally well. Some remain in one place, but others are wandering; and some make long periodical journeys or migrations. At the time for spawning or laying their eggs, fishes in the sea generally either approach the coasts or go up the rivers. The Herrings are an example of the former, and the Shad and Salmon of the latter. In these migrations the Salmon observe regular order, as the wild geese do in theirs. They form two long files, united together in front, and led by the largest female in the troop. The males form the rear guard. When any obstacle opposes, they leap over it, sometimes to the height of ten or even sixteen feet. In this way they ascend rivers nearly to their sources, and deposit their eggs in the autumn in holes which they dig in the sand. Remaining here through the winter, in the early spring they return to the sea. It seems that the Salmon have the same instinct that some birds have in regard to place, § 212. This was proved by a naturalist named Deslandes in this way. He placed a ring of copper on twelve of these fish, and set them at liberty in the River Auzou, in Brittany. They, of course, emigrated, but the next year five of them were caught in the same place, the second year three, and the year after three more.

363. Most fishes are abundantly prolific. You can see this to be so if you observe the roe or spawn of any fish, this being the collection of the eggs of the animal. It is estimated that at least 60,000 eggs are contained in the roe of a Herring. The roe of a Codfish was ascertained

to contain nine million of eggs. Fishes being thus prolific, societies of them, or shoals, as they are called, are often immense in multitude. They would be too abundant were their number not kept down by various causes. Many of the eggs are destroyed, and then of the young fish so many are eaten by other fish, and are killed in various ways, that few of them comparatively come to maturity.

364. Some fishes present a strong contrast to all this in the number of their young. This is the case with that rapacious fish, the Shark, thus illustrating the Divine wisdom and benevolence. It produces but two eggs. The eggs of some species of the Sharks are great curiosities. They are of firm texture, and of a purse-like shape, with a long tendril extending from each corner of it, as seen in Fig. 168. These tendrils, coiling around seaweed or any

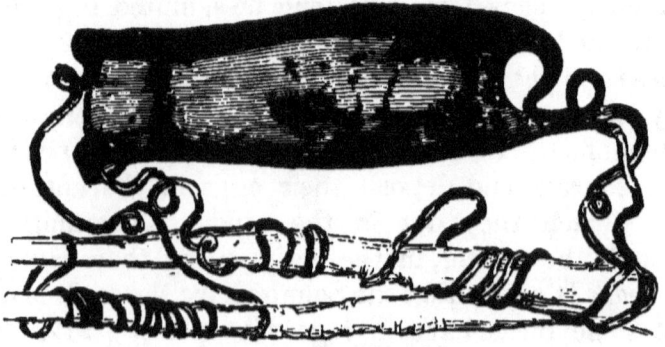

Fig. 168.—Egg of Shark.

other substance, serve to anchor the egg securely. The purse is thin at the end where the head of the young fish is, and when it is in a fit state to come out, it breaks its way through this end. Some other fishes lay similar eggs. They are sometimes picked up by the sea-shore, and are called Mermaids' Purses.

365. Fishes supply quite a large portion of the food of the human family. An immense amount of capital is employed in carrying on the fisheries, and in some quarters a large part of the population are engaged in them.

It was estimated at one time that one fifth of the population of Holland were devoted to this branch of industry alone.

366. There have been various systems of classification proposed for the Fishes. Cuvier first divides them into those that have really bony skeletons, and those that have cartilaginous ones. He then divides the Osseous or bony fishes into two groups according to their fins, the first being spine-rayed, the second soft-rayed. The Cartilaginous fishes he divides into two groups according to the arrangement of their gills, the fringes being free in the one, and being fixed in the other. Professor Agassiz classifies fishes according to the character of their scales, making four orders.

367. It would take us into too broad a field to go into the minute classification of fishes. I shall, therefore, in addition to what has already been presented, notice particularly only a few of the most interesting of these animals.

368. The Swordfish, Fig. 169, is found in every part

Fig. 169.—The Swordfish.

of the Mediterranean Sea. Its "sword" is an elongation of the upper jaw, of great strength. It uses it in transfixing its prey, running into shoals of fishes for this purpose. In the British Museum there is a piece of the bottom of a ship with a "sword" thrust entirely through it. The length of this fish is from twelve to fifteen feet. Another fish of about the same size has a similar projection from the upper jaw, but notched on both sides, and hence it is called the Sawfish. With this instrument it sometimes attacks the Whale, inflicting severe wounds on him, and sometimes imbedding the saw in his body in its full length.

369. The John Dory, Fig. 170, is a singular fish in its

Fig. 170.—The John Dory.

shape, its markings, and its appendages. In England its fame is associated with the performances of Quin the comedian. There are various traditions of a curious nature in regard to the round spots on its sides. One is, that this is the fish that St. Peter caught, and that in taking the tribute-money out of its mouth he made these marks with his finger and thumb. The name of this fish is probably a corruption of the French *jaune doré*— golden yellow, the color of the lighter parts of the fish when it is alive.

Fig. 171.—The Seahorse.

370. The Seahorse, Fig. 171, has been

often found off the coasts of England. It is the only fish that is as yet known to have a prehensile tail. It has been found in the Hudson River of this country, about five or six inches in length.

371. **The Lophius, or Fishing Frog**, Fig. 172, appears

Fig. 172.—The Lophius, or Fishing Frog.

on all the European coasts, and also on our own. With its pectoral fins it can crawl on land. The voracity of this fish is very great, and if caught in a net with other fish it will devour some of its fellow-prisoners. Its usual mode of capturing its prey is this. Crouching close to the bottom, and stirring up the mud and sand, it moves about the long filaments; the small fishes, swimming about, suppose these filaments to be worms, and as they are about to seize them, the Angler, with a quick movement, takes them into his capacious jaws.

372. **The Sturgeon**, Fig. 173 (p. 220), although one of the cartilaginous fishes, has externally rows of bony plates. It is very common in the northern parts of Europe, where there are regular fisheries for its capture. Almost every part of it is used—isinglass being obtained from its air-bladder, and caviar from the roe, while the flesh is consumed both in a fresh and a salted state. It is much esteemed as food, eaten fresh, in the Atlantic

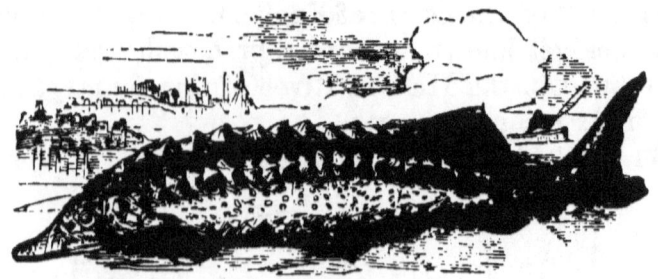

Fig. 173.—The Sturgeon.

Southern States of this country. The female deposits her eggs in the fresh water of rivers, and the young, when hatched, migrate to the sea.

373. The family of Flat Fish differ in some important respects not only from all other fishes, but from all other vertebrated animals. The Turbot, Fig. 174, is an ex-

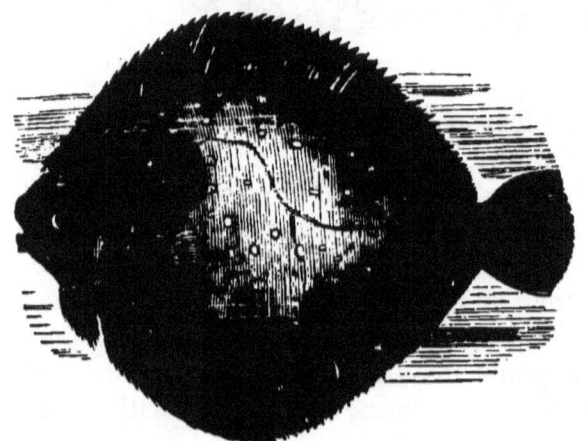

Fig. 174.—The Turbot.

ample of it. These fishes are not ordinarily in the position that you see here, but lie flat along near the bottom, the upper surface in most species being of a dark color, while the lower surface is white. They have no air bladder, and have little power of rising in the water. When they are disturbed they assume a vertical position, showing their white sides, and dart along with great rapidity. Some are occasionally met with in which the sides are alike in color, most often dark, but sometimes white, and these are said to be "doubles." The fishes of this family differ from all other Vertebrates in a lack of perfect symmetry in relation to the two halves of the body. In all other Vertebrates the two halves are alike, except in some of the internal organs. But in these fishes there is a difference between the two sides in several particulars. The two eyes are on one side, and are irregularly placed; in some species they are on the right side, and in others on the left. With this there is an irregularity in the bones of the head, while in all other Vertebrates they are alike on the two sides. Then there is the difference in color. There is also a difference in the pectoral fins, one being longer than the other.

374. Most of the species of this family belong exclusively to the sea, and yet the Flounder and some others occasionally ascend rivers. The Halibut, so well-known in this country, is a very large fish, sometimes six or seven feet long, and weighing four or five hundred pounds. The flesh of the Turbot is considered peculiarly fine, and immense quantities are taken in the fisheries. Carpenter states that the Dutch receive £80,000 per annum for the supply of this fish to the London market.

375. The Herring family furnish a great amount of food to man. They are mostly marine fishes; only a few species, among which is the Shad, ascending the rivers at the spawning season. The common Shad of this country has a much better flavor than the Shad of Europe. The Sardine and Anchovy are aberrant species of this family. The true Herring inhabits the northern seas, and arrives every year in vast legions on the coasts of America, Europe, and Asia, never descending below the 45th degree of north latitude. They come to the coast to spawn, and then retire to the depths of the sea, going northward.

376. The Eels are called Apoda, or Footless Fish, be

cause they have no ventral or belly fins. They have long, snake-like bodies, covered with a soft skin, the scales being very minute, often almost invisible. They can live for some time out of the water, chiefly from a peculiar arrangement of the gills. The gills have very narrow openings, and are, therefore, so much sheltered from the air that they do not readily become unfit for respiration in becoming dry (§ 350). There is a similar arrangement in the Lampreys, a class of Fishes of eel-like shape, in some respects the lowest in organization of all the Vertebrate animals. In these fishes there are fourteen gill-openings, seven on each side, as seen in Fig. 175.

Fig. 175.—Lamprey.

They are sometimes called Seven-eyes on this account. The mouth is a singular apparatus. It is ring-shaped, and is armed with numerous teeth, and there are also two longitudinal rows of small teeth on the tongue. The tongue moves backward and forward in the mouth, acting as a piston, thus, by its suction power, enabling the fish to hold on to any object that it pleases.

377. In the rivers and ponds of Surinam and other parts of South America there is found an Eel which is armed with a true electric battery. It uses it in destroying the life of its prey, which it does instantaneously. It can sometimes give a shock powerful enough to prostrate a man. Humboldt describes the method adopted by the natives in taking these animals. Having found a pool in which they are, they drive in a troop of wild horses. Aft-

er the electricity accumulated in their batteries is pretty much expended on the horses—some of them, perhaps, being killed by it—the Eels are captured with impunity.

378. There is an aberrant genus of the Ray family which has a similar apparatus, the situation of which is seen in Fig. 176, in the two elevations extending from the

Fig. 176.—Torpedo.

eyes about half way down the body. This fish is found chiefly in the Mediterranean, where its powers are well known and are much feared. The apparatus is represented in Fig. 177 (p. 224), the batteries on each side being at *e*. On one side is seen the nerve, which, branching out from the brain, *c*, to the battery, is the means by which the animal can work it at pleasure. The batteries are composed of multitudes of tubes pressed one against another like the cells of a honeycomb, and filled with a thick fluid. The true Rays have on the tail and other parts barbs or prickles with which they can inflict wounds. They are shaped much like the Flatfish (§ 373). But in their case the upper side is really the back and the under side is the belly; and they are symmetrical, having the eyes on the upper side, and the mouth, nostrils, and gill-

224   NATURAL HISTORY.

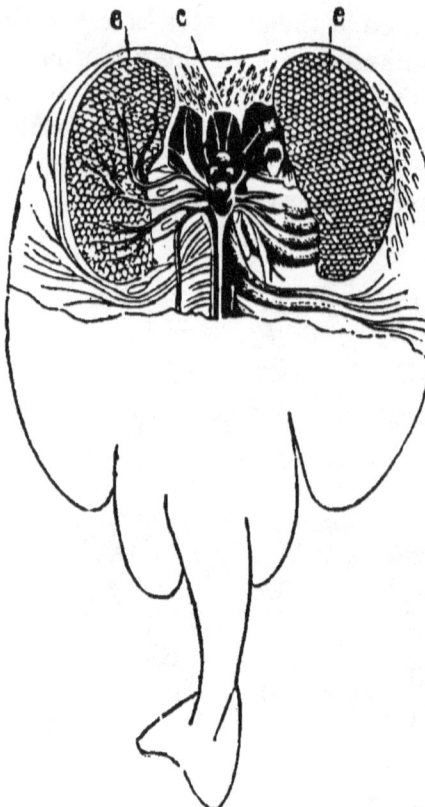

Fig. 177.—Electric Apparatus of Torpedo.

openings beneath. Like the Flatfish, they move along in search of their prey near the bottom. Some species reach a large size, the Eagle Rays having been seen twenty-five feet long and thirty broad. It is stated that one was taken at Barbadoes which weighed 3500 pounds, and that it required seven yoke of oxen to draw it ashore.

Questions.—What is said of the distinction between salt-water and fresh-water Fishes? What is said of the migration of Fishes? What of the migration of the Salmon? Give the observation of Deslandes. What is said of their fecundity? What is said of the Sharks in this respect? Describe the eggs of some Sharks. What is said of Fishes as supplying man with food? What is said of the classification of Fishes? What is said of the Swordfish? Of the John Dory? Of the Seahorse? Of the Lophius? Of the Sturgeon? What are the characteristics of the Flatfishes? How do they differ from all other Vertebrates? In what water are most species found? What is said of the Halibut? What are some of the fishes of the Herring family? What is said of the true Herrings? Why are Eels called Apoda? What is said of their scales? What of their gills? What of the gills of the Lampreys? What of their mouth? What is said of the Electrical Eel? Describe the arrangement of the electrical apparatus of the Torpedo. What is said of the true Rays? Give the comparison with the Flat fish. What is said of the size of the Rays?

CHAPTER XXII.

INSECTS.

379. The sub-kingdom which we have already considered is the Vertebrate. The other three sub-kingdoms are said to be Invertebrate, the prefix *in* being here used as meaning the same as the very common negative prefix *un*. Of these sub-kingdoms I will notice the Articulata first. The two chief characteristics of it were stated in Chapter I. These, however, and other characteristics, require a more particular notice here, before I enter on the consideration of Insects, the special subject of this chapter.

380. The kingdom of the Articulates includes a wide range of animals of great variety—Insects, Worms, the Spider and Scorpion tribe, and the Crab tribe. But they all agree in one thing—in having a covering which answers the purpose of the internal skeleton of the Vertebrates. This covering, or jointed armor, gives firmness to the body, and furnishes points of attachment to the muscles.

381. This skeleton coat of mail is very commonly arranged on the bodies of these animals in segments in the form of rings. This arrangement is seen most perfectly carried out in the Centipede, Fig. 178 (p. 226). You can also see it plainly in the bodies of most insects, as, for example, in the common Fly. This arrangement mostly disappears in the Crab tribe, where, for the sake of firmness, the skeleton covering the body of the animal is in one piece. So, also, on the other hand, it nearly disappears in such soft Articulates as the Leech and Earthworm, because there the rings would not allow the requisite limberness.

382. I have spoken in § 15 of the ganglia of the nervous system of the Articulates. Commonly there is a chain of them, as seen in Fig. 8. But sometimes, as in some of the Crab tribe, there are only two, one in the head and the other in the thorax. This is a decided approach toward the arrangement of the nervous system of the Vertebrates, the upper ganglion being somewhat like a true brain.

383. The muscles constitute the bulk of the body in the Articulates; and, being thus muscular, they are very active. For this purpose the armor-skeleton is made as light as possible consistently with the requisite firmness. The movements of some of them exceed in rapidity those of any other animals. With the exception of one group, they roam freely in search of their food, and have very effective organs for capturing their prey. Their apparatus for mastication, also, is commonly complicated and powerful.

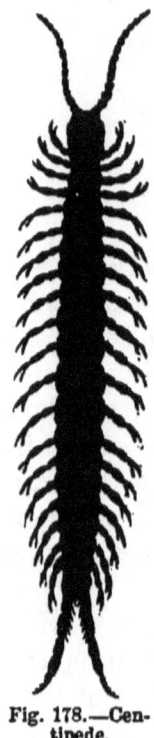

Fig. 178.—Centipede.

384. Almost all of the Articulata have a distinct head. The jaws do not move up and down as in the Vertebrates, but sidewise. There are often several pairs of them, sometimes having cutting edges, sometimes edges with saw-like teeth, and sometimes they are fitted to crush rather than to cut or tear. The legs are generally six, eight, ten, or fourteen in number, but sometimes there are many hundreds. Sometimes there are none; but when they exist at all, they are never less than six.

385. The circulation of the Articulates is peculiar. There is no heart, but instead there is a tube stretching along the back. This is not perfectly regular, but has segments corresponding with the segments of the body mentioned in § 381. Each segment is a sort of heart for

its division of the body. The blood is for the most part white.

386. The respiration differs in the different classes. In the Crustacea, the class to which the Lobsters and Crabs belong, the respiratory organs are, like those of the Fishes, fitted for the action of air mingled with water. They are really gills in various forms. But in the Insects and Spiders the respiration is aerial; that is, the respiratory organs are fitted for the action of air alone by itself.

387. The symmetry of the body is even more complete in the Articulates than it is in the Vertebrates. In the latter there are many of the internal organs in regard to which the two halves of the body are not alike, as the heart, stomach, lungs, etc. But in the Articulates this symmetry is extended even to some of these internal organs.

388. The Articulata are commonly divided into seven classes. 1. Insects. These have the three divisions of the body, the head, the thorax or chest, and the abdomen. They have antennæ or feelers on the head, three pairs of legs, and generally one or two pairs of wings. 2. Myriapoda, the Centipedes. They have no separation of the body into thorax and abdomen. The head, however, is very distinct. There are seldom less than twenty four pairs of legs. The segments of the body are numerous and equal. 3. Arachnida, including the Spiders, Scorpions, and Mites, are characterized by the union of the head and thorax, by the very distinct separation of this cephalo-thorax from the abdomen, by the possession of four pairs of legs, and by their want of antennæ. In all these three classes the respiration is aerial. 4. Crustacea, or Crabs, Lobsters, etc. The respiration is aquatic, like that of Fishes. They have from five to seven pairs of legs. The body is variously divided. In some the three parts are distinct, as in Insects; in others the arrangement is like that of the Arachnida; and in others still it is like that of the Myriapoda. The classes already

mentioned have articulated members. The remaining three have none. 5. The Annelida, the Leech and Worm tribe, having the segments very indistinct. 6. Entozoa, Intestinal Worms, in which the *articulated* arrangement is still more indistinct than in the Annelida. 7. Rotifera, or Wheel Animalcules, very minute animals, presenting the articulated character quite indistinctly.

389. I now pass to the consideration of Insects. This class of the Articulata has an immense number of distinct species, surpassing in this respect every other department of the animal kingdom. Dr. Carpenter estimates the sub-kingdom of Vertebrates as containing 30,000 species, a number which is exceeded by one single order of the Insect tribe, the Beetles. And numerous as are the known species of Insects, it is supposed that the number of those which remain to be discovered is far greater.

390. The name Insect comes from the Latin word *inseco*, to cut into, and refers to the divisions or sections of the animal's body. The intervals between them are so abrupt that they appear as if made by a cutting operation. The sections or segments are usually thirteen or fourteen in number. One is the head; the chest or thorax has three; and the abdomen nine.

391. The respiration of Insects is peculiar. They have no lungs in one particular part of the body. Their lungs may be said to be in all quarters of the body, for air is admitted by various openings into tubes which traverse here and there. It is thus that the blood of the insect is acted upon by the air. These openings are generally mere slits like button-holes; but often they have two valves which open and shut like folding doors, and sometimes they have over them a sort of fine grating to keep dust from entering. As Insects are exceedingly active, they require comparatively a large amount of air. They are strikingly in contrast with Reptiles in this respect. These latter are so dull and slow that they need but lit-

tle air; and when they go to sleep for the winter they require none, and their breathing stops.

392. Some insects live on the juices of plants and of animals, and some devour the substance of either the one or the other. The former suck their food; the latter gnaw it. These two classes, therefore, have two different kinds of mouths. The gnawers, such as Beetles, Cockroaches, Locusts, etc., have a complicated apparatus, which I will describe. First, there are two tooth or claw-like appendages, called mandibles; these are the upper jaws, which divide the food. They come together by a lateral or sidewise motion. Sometimes they have sharp edges to cut like scissors, and sometimes they have points for tearing. Below or behind these are two other jaws, called *maxillæ*, which are very complex in their structure. Above, or, rather, in front of the mandibles, is a lip, and so there is one behind the maxillæ. Insects furnished with an apparatus of this kind are called *mandibulate*.

393. In some insects we have an arrangement entirely of a different character, as in the Butterfly tribe. Here there is a tubular appendage, or trunk, often quite long. This is ordinarily coiled up, as you see in Fig. 179. When the animal wishes, it can uncoil it and extend it down into the bosom of flowers. Such insects are called Haustellate, from *haustellum*, a sucker. This tube, or proboscis, varies much in different insects. In some, as the Bees, there is a mixture of the mandibulate and the haustellate arrangements. They obtain their food by suction, and use their mandibles and maxillæ as trowels and spades, and knives and scissors, in building their curious habitations. In insects that suck the blood of animals, such as the Musquito and the Horsefly, there is a peculiar modification of the apparatus. There is a proboscis with lancets to make

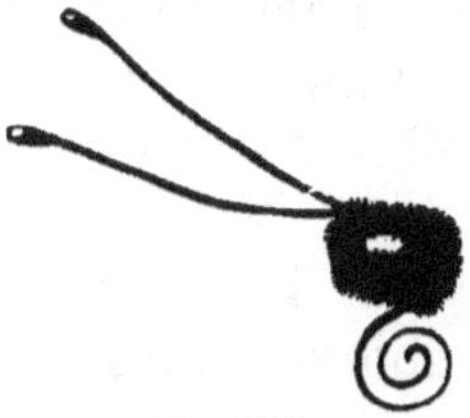

Fig. 179.

the necessary wound in the skin of the animal whose blood is to be sucked.

394. The head of the insect is furnished with certain appendages called *antennæ*, supposed to be organs of feeling, and perhaps of hearing also. They are various in form, and commonly very beautiful, especially when examined with the microscope. In Fig. 180 you have

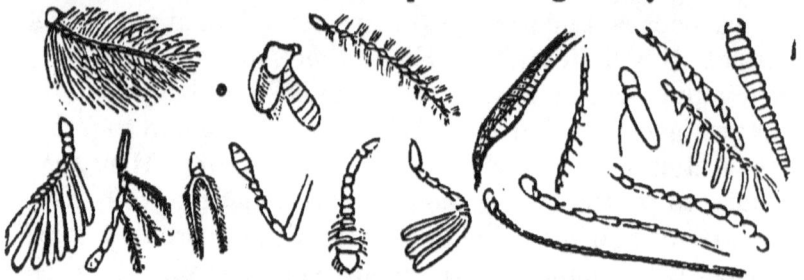

Fig. 180.—Variously-formed Antennæ of Insects.

a variety of them represented. There are other feelers called *palpi*, which are usually much smaller and shorter. The antennæ probably act as feelers in regard to objects a little distance off, while the palpi do the same for substances close by the mouth of the animal.

395. The senses of Insects are all acute; and yet the organ of smell has never been discovered, and in most no organ of hearing can be found. The organs of vision are generally plain to be seen, and are often exceedingly prominent objects as we look at the insect. Very commonly the eye is a multitude of small eyes. In the common house Fly there are 4000 of them, and some insects

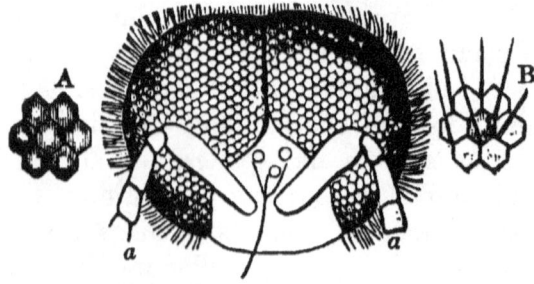

Fig. 181.—Head and Eyes of the Bee, showing the Division into Facets.

have 25,000. If examined with a microscope, the surface of these compound eyes appears as you see in Fig. 181 (p. 230), representing the head and eyes of a bee. At *a* are the antennæ, and the eyes occupy most of the front and sides of the head. The surface is made up of hexagonal (six-sided) facets. At A you see them much magnified, and at B you see hairs standing out between them. Each of these facets is a cornea, or window to a little eye. (See "First Book in Physiology," p. 167.) Most insects have these compound eyes, but some have only single ones, and some have both, for what reason we do not know.

396. The digestive apparatus is commonly quite complicated, there being three stomachs—one corresponding to the crop of birds, another to the gizzard, while a third receives the food and digests it after it has been softened and ground in the other two stomachs. The second stomach, or gizzard, is often armed with horny projections, in order to grind up the food effectually. In Fig. 182 you have the whole digestive apparatus of a Beetle. First you see the strong cutting mandibles, the maxillæ and the antennæ; then *a* the throat, *b* the gullet, *c* the crop, *d* the gizzard, *e* the third stomach, *f* the intestine. The liver, *h*, instead of being such a solid organ as it is in vertebrate and most other animals, is

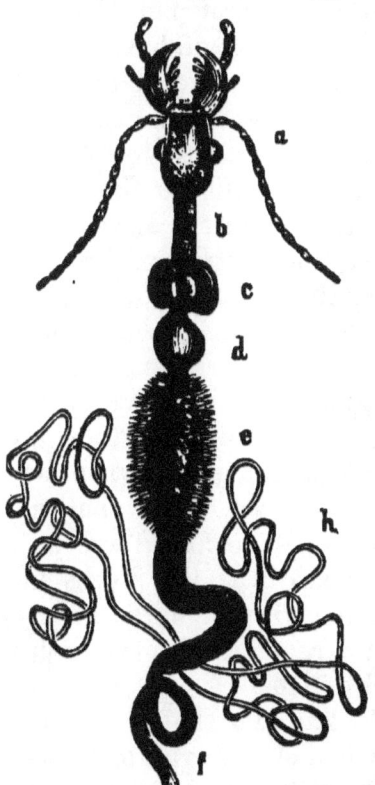

Fig. 182.—Digestive Apparatus of Beetle.

made up of long and delicate tubes. As lightness is a great object in the structure of the Insect, the digestive apparatus is made of as little bulk as possible.

397. The feet of Insects are in conformity with their modes of life. Some have claws or hooks; some have a kind of suction cushion by which they can adhere to surfaces; some have fringed feet to enable them to swim; and some have their fore feet shaped for digging, like the Mole's.

398. The wings are generally made very much like those of the Bat, § 58. They consist of a double membrane over an extended slender frame-work. There are generally two pairs, but sometimes only one, as in the case of the common Fly. Often the first pair of wings are mere coverings for the other wings, and have no active agency in flight. In this case they are made thick and firm, and are called the *elytra* (singular elytrum). In Fig. 183 you see the elytra at *a*. When the Insect is at rest, the elytra are brought together over the back, the true wings being folded, sometimes very curiously, under them. The true wings are light and thin, and they are transparent, except when they are covered with what appears to the naked eye as a kind of colored dust, as is the case with the Moths and Butterflies. This dust, examined with the microscope, is found to be made up of little regularly-formed scales, often beautifully marked with lines. When they are rubbed off, their fastenings look, under the microscope, like the nail-heads on a roof when old shingles have been torn off. In some of the Butterflies the scales are arranged like shingles on a roof, and with their various colors present a very beautiful appearance.

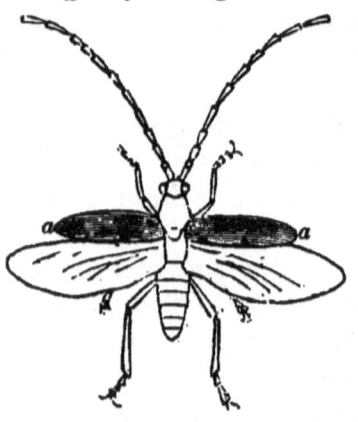

Fig. 183.

399. There is no part of the animal world that exhibits so great a variety of beauty in both color and structure as the Insect tribe. And it is especially true of these animals, that, great as is the beauty visible to the naked eye, still greater is that which the microscope reveals.

400. With very few exceptions Insects are oviparous. In many cases the eggs are laid in autumn, and are hatched in the spring. Insects that do this, instinctively make special provision for the preservation of the eggs through the winter. When the eggs are of a tender consistence, the Insects deposit them deeply in the earth. But some are deposited on trees. In this case leaves are not selected as the place of deposit, as is very commonly done when the eggs are to be soon hatched; for if this were done, the leaves being scattered by the wind, the Insects, when they came to be hatched, might be far away from their appropriate food. Accordingly, the parent deposits the eggs on the trunks or branches of the trees, upon whose young leaves their progeny can live when the warmth of spring hatches them. Moreover, eggs thus exposed to the cold of winter have not the usual delicate covering of those which are hatched the same season in which they are laid, or of those which are deposited in the earth for the winter. They are covered with a hard, thick shell; and, besides, they are well packed together, and the interstices are filled up with a tenacious substance which becomes very hard. The arrangement of these little eggs is often very beautiful. They are of very great variety of shape, and some of them are curiously and elaborately constructed. In Fig. 184 (p. 234) you have a few of these varieties represented.

401. Insects are commonly exceedingly prolific. The queen of the Honey Bees lays fifty thousand eggs, and the female White Ant produces forty or fifty millions in a year. It is calculated that the progeny of a single Aphis or Plant Louse number in one season a trillion,

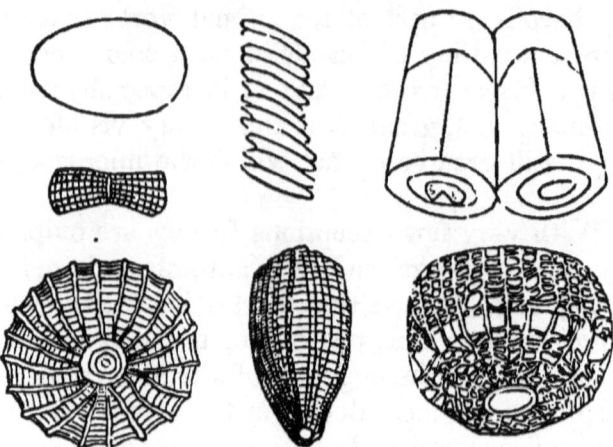

Fig. 184.—Magnified Eggs of various Insects.

that is, 1,000,000,000,000,000,000. But they are so feebly constructed that a large portion of them are destroyed in one way and another before they come to maturity. Insects are distributed largely over all parts of the globe, appearing even in the arctic regions during their short summers. They are the most abundant in the tropical regions, and there the largest and most brilliant species are found. Each region has Insects peculiar to itself. Some, however, are very widely distributed, the common House Fly the most widely of all.

Questions.—What is the meaning of the term Invertebrate? What are included in the sub-kingdom of the Articulata? In what are they alike? What is said of the arrangement of the skeleton-covering of the Articulata? What is said of their nervous system? What of their muscles? What of their jaws? What of their circulation? What of their respiratory organs? How do they differ from the Vertebrates in symmetry? Give the classes of the Articulata, with their characteristics. What is said of the number of Insects? Show the appropriateness of the name Insect. Describe the respiratory apparatus of Insects. What do Insects live on? Describe the mandibulate apparatus. Describe the haustellate apparatus. What are some of its variations? What is said of the antennæ? What of the palpi? What is said of the senses of Insects? Describe the arrangement of the eyes of a Bee. What is said of the digestive organs of Insects

Describe the digestive apparatus of a Beetle. How are the wings of Insects constructed? What are elytra? What is there peculiar in the wings of Moths and Butterflies? What is said of the beauties of Insects? What various provisions are made for preserving and hatching the eggs of Insects? What is said of their shapes? What is said of the fecundity of Insects? What of their distribution?

CHAPTER XXIII.

THE METAMORPHOSIS OF INSECTS.

402. THE grand peculiarity of Insects is their metamorphosis, or change of form. Almost every Insect undergoes this change, there being commonly three distinct changes of being. In the first stage the Insect is a crawling caterpillar or a worm. In its second stage it is wrapped up in a covering prepared for the purpose, and is in a state of sleep. During this sleep great changes are going on. When these are completed it is a winged animal, its wings being closely folded up. In due time it comes out of its prison, and spreads its wings for flight. It is now deemed to have arrived at its perfect condition.

403. In its first stage it is called a *Larva*, this being the Latin word for mask, the idea being that the Insect is now not in its true state or character, but is in a masked condition, from which it will after a while come out. When it does so, it is called the *Imago*, or said to be in the imago state. The Insect is now the image or representative in full of its species. Its sleeping state, the one intermediate between the larva state and the imago state, is a transition one. In this the Insect is changing from a crawling to a flying animal. It is now termed a *Pupa*, the Latin for baby, because it commonly appears somewhat like an infant trussed up with bandages, as has sometimes been the fashion in some nations. The or-

dinary forms and appearance of Pupæ (plural of Pupa) are represented in Fig. 185.

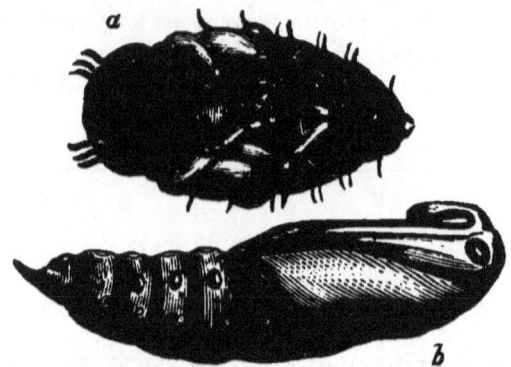

Fig. 185.—*a*, Pupa of a Water-beetle (*Hydrophilus*); *b*, Pupa of *Sphinx Ligustri*.

404. The different larvæ of Insects have the different names of maggot, grub, and caterpillar, according to their form and appearance. The pupæ of Butterflies and Moths were formerly called Chrysalids and Aurelias, because the coverings of some of them have spots of a golden hue. The term Chrysalis is often used at the present day as synonymous with pupa, and this state of the Insect is called the Chrysalid state.

405. The changes which take place in the pupa state are very great, even radical ones. There is commonly no resemblance between the Larva and its Imago. There may be great beauty in the Imago, and none in the Larva, and sometimes the reverse is the case. Then, as to form and general structure, the contrast is of the most marked character. In the Larva state it was a slow, crawling animal, but in the Imago state it is light, perhaps delicate in structure, and is nimble on the wing. And the change is as great internally as it is externally. Its stomach even is changed, for its mode of getting a livelihood is different now. There are corresponding changes also about the mouth, a coiled tongue perhaps appearing in place of the formidable gnawing apparatus of the larva. In relation to this change it has been well

said, "Were a naturalist to announce to the world the discovery of an animal which for the first five years of its life existed in the form of a serpent; which then, penetrating into the earth, and wearing a shroud of pure silk of the finest texture, contracted itself within this covering into a body without external mouth or limbs, and resembling more than any thing else an Egyptian mummy; and which, lastly, after remaining in this state for three years longer, should, at the end of that period, burst its cerements, struggle through its earthy covering, and start into day a winged bird, what would be the sensation excited by this piece of intelligence?" And yet this would be no more wonderful than the ordinary metamorphosis of Insects. Indeed, many of the most marvelous circumstances in this change are not at all referred to in the supposition above made.

406. The larva is produced from an egg, and the egg is laid by the perfect Insect or Imago. When the larva is first hatched it is very small, but it grows with a rapidity always great, in some cases enormous. The maggots of flesh Flies are said to increase in weight two hundred times in twenty-four hours. To make such an increase these animals must eat voraciously. With the great multiplication of their number, the amount which a collection of them will sometimes devour is wonderful. Linnæus calculated that three flesh Flies and their immediate progeny would eat up the carcass of a horse sooner than a lion would do it.

407. In the Imago state the Insect eats but little, as it grows little or none ordinarily. The Butterfly or Moth comes forth from its prison fully grown; but the caterpillar from which it was formed was very small at the outset, and became large by large eating. Our common Flies are small and delicate eaters, but the Maggots, the larvæ from which they came, rioting in filth, devour largely what the Flies will not touch.

408. The great growth of larvæ obliges them to cast

their skins repeatedly. The Silkworm and other caterpillars cast their skins about four times during their growth.

409. Insects pass the time of their pupa state under various circumstances. Some, when about going into this state, crawl into some by-place away from intruders. Some work their way into the ground, and perhaps spin a silken lining for the earth-cells in which they are to sleep through their change. Some roll themselves up in leaves. Some construct for themselves a silken house, called a cocoon, attached to some leaf or twig.

410. Among those that do this last is the Silkworm. The formation of the cocoon I will describe. When the worm has its silk factory, which is near its mouth, properly stocked with the gummy pulp from which the silk is to be spun, it seeks a good place where it can have a sort of scaffolding for its cocoon. It first spins some loose floss, attaching it to things around. Next it begins to wind its silk round and round, making a cocoon at length, shaped much like a pigeon's egg, being smaller at one end than the other. It thus gradually shuts itself up in a silken prison. The last of the silk which it spins is the most delicate of all, and it is well glued together, making a very smooth surface next to the Silkworm's body. The silken house being constructed, it now prepares itself for its sleep and its change. It sheds its skin now for the fourth and last time, tucking its old clothes, as we may say, very snugly at one end of the cocoon. It then passes into its sleep, and a new and thin skin is formed over it, in which it gradually changes into an animal endowed with wings. At the proper time it works its way out of its prison, unfolds its wings, and flies off, not to eat mulberry leaves, as it did in its larva state, but to sip the honey from the flowers.

411. Observe the manner of its exit and the arrangements for it. The head is always at the small end of the cocoon, and here the silk is less closely wound and less

tightly cemented by the gluey substance. The old clothes are always at the other end, so as not to be in the way. The new coat which was formed as it entered the pupa state is easily torn, and the Moth, moistening the cocoon with a fluid from its mouth at the part where it is to escape, easily forces its way through. The opening from which it emerges is very small, and the shape of the animal before it expands its wings is that of a long bundle.

412. The thread with which the worm makes its cocoon is an unbroken one. It can, therefore, be unwound or reeled off, which is done in obtaining it for manufacture. For this purpose the cocoons are exposed to the heat of an oven in order to kill the pupæ in them, and then, by a little soaking in warm water, the glutinous matter which unites the silk is so softened that the thread can be readily unwound. The length of it varies from six hundred to a thousand feet; and as it is double as spun out by the insect, its real length is nearly two thousand feet. So fine is this double thread, that the silk that comes from one cocoon does not weigh above three and a half grains, and it requires ten thousand cocoons to supply five pounds of silk. The native countries of the Silkworm are China and the East Indies; and in ancient times the manufacture of silk was confined to them. So scarce was the article in other countries, even as late as James I. of England, that this monarch, before his accession to the throne, wore on some public occasion a borrowed pair of silk stockings. But at the present time the culture of the Silkworm and the manufacture of silk are so widely diffused, that silk is every where, in civilized communities, one of the common articles of dress.

413. When a pupa is to remain out of doors all the winter, special pains are taken to guard it against the cold. For this purpose great numbers of Insects in the autumn dig their way down into the ground, and pass their pupa state in an earthy cell below the reach of

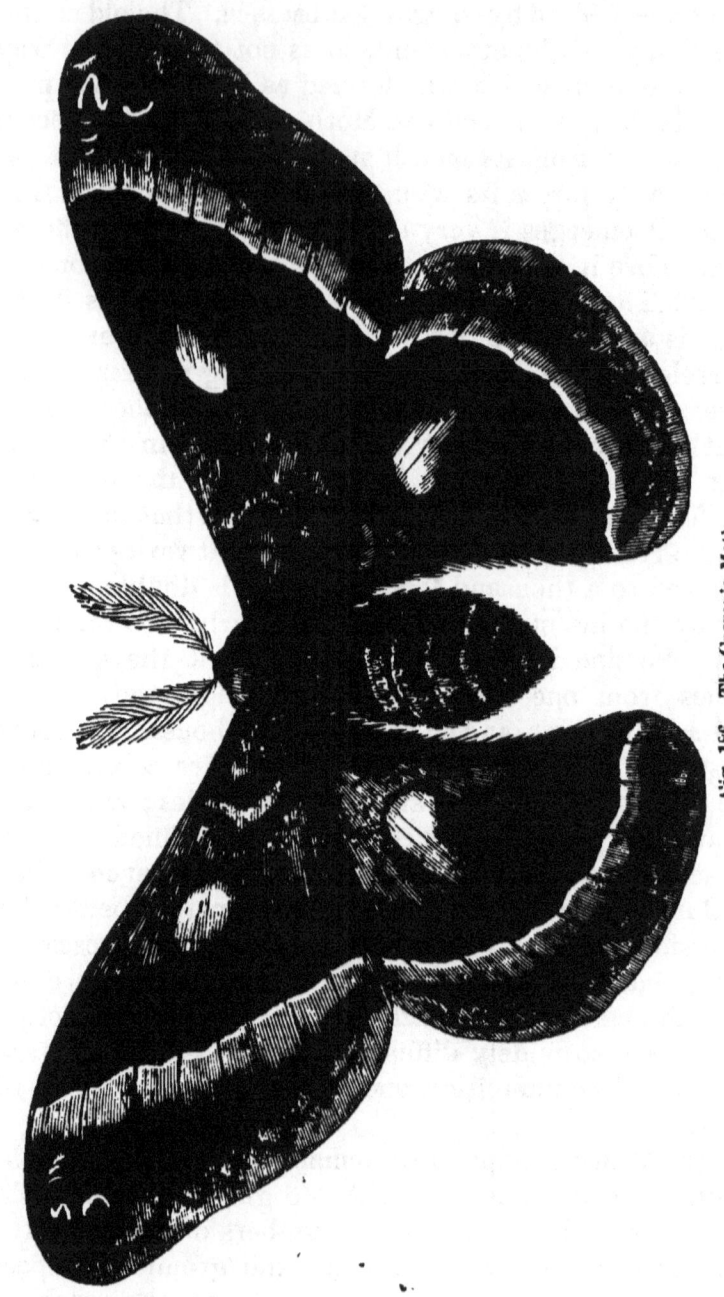

Fig. 156.—The Cecropia Moth.

frost. Some line this cell with silk, making thus a soft covering for the body, and shutting out more effectually the cold. Some of the caterpillars accomplish the same object by constructing above ground a cocoon specially adapted to guard against the cold. This is exemplified in the case of one of the largest and most splendid of our American Moths—the Cecropia Moth, Fig. 186 (p. 240). It is found, as Professor Jaeger states, all the way from the Canadas to the Mexican Gulf, and also in all the Western States. It has large wings, measuring five to six inches from tip to tip. The scales on them, § 397, are dusky brown. The borders of the wings are richly variegated, the anterior ones having near their tops a dark spot resembling an eye, and both pairs having kidney-shaped red spots.

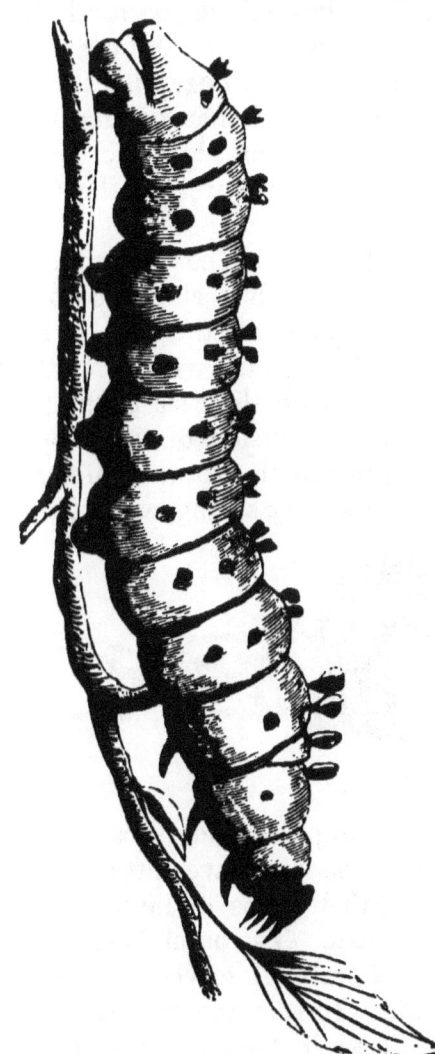

Fig. 187.—The Caterpillar, or Larva.

414. In this case the caterpillar, or larva, Fig. 187, is nearly as beautiful in colors as the perfect insect or imago. It is of a light green color, and has coral-red warts, with short black bristles, over its body. It feeds

on the leaves of trees till August or September, and then descends to seek for some currant or barberry bush upon which it may build its house for its winter sleep. "Any one," says Professor Jaeger, "who meets with these caterpillars in the above-mentioned months may have the pleasure of witnessing their metamorphosis into cocoons, and several months after into an elegant moth, by taking them up very carefully upon leaves and carrying them home, placing them in a spacious box, with a little undisturbed earth at the bottom, and then putting into it some dry brush-wood, about one foot high, and covering the whole with gauze in order to prevent their escape."

415. I will now describe the peculiar construction of the cocoon. That of the Silkworm is a simple cocoon, no special provision being made against the cold, as the pupa state, instead of lasting through the winter months, is finished in a few weeks. But in the case of the Cecropia Moth there is a covering outside of the proper cocoon. This covering is fastened to a branch of some bush, as in Fig. 188. It is made very strong, as its fibres

Fig. 188.

are much more closely joined together than those of the cocoon inside of it. Often there are leaves attached to it, leaving the impression of their veins or nerves upon it when you have detached them. The animal evidently uses these leaves as a sort of scaffolding when it begins to construct its winter home. In spinning this covering it works all the while inside, as it does in spinning the cocoon. After finishing it, it lines it with coarse loose silk, and then proceeds to spin its cocoon in the same way that the Silkworm does, making it of the same shape. The loose silk between the cocoon and the outer covering is blanketing for the purpose of warmth. By these means the pupa or chrysalis is secured against dampness and cold, and amid all the storms of winter is even more safe from harm than an infant in its cradle under the watch of an anxious mother.

416. As in the case of the Silkworm Moth, the Cecropia always comes out at the smaller end, and here both the cocoon and the outer covering are made less close and strong than in the other portions. In New England this Moth comes forth in June. Last year I obtained from my garden two cocoons which were near each other on a currant bush. I gave one to a lad living on Staten Island, and retained the other myself. His Moth came out three weeks before mine, corresponding with the advance of the season there before ours. When mine emerged I caught the same evening in my house two others, and on the following evening three more. As we saw none before or after, this seems to show that these Moths come forth almost simultaneously in the same locality.

417. Dr. Harris, in his work on the Insects of New England, recommends a trial of the manufacture of silk from the cocoons of the Cecropia and some other of our large indigenous Moths. "Their large cocoons," he says, "consisting entirely of silk, the fibres of which far surpass those of the Silkworm in strength, might be employed in the formation of fabrics similar to those manufactured in India from the cocoons of the Tusseh and Arindi Silkworms, the durability of which is such that a garment of Tusseh silk is scarcely worn out in the lifetime of one person, but often descends from mother to daughter; and even the covers of palanquins made of it, though exposed to the influence of the weather, last many years. Experiments have been made with

Fig. 189.—Silk of the Cecropia.

the silk of the Cecropia, which has been carded and spun, and woven into stockings that wash like linen." The silk can be very readily reeled off from the cocoons as seen in Fig. 189.

418. Some Insects go through an imperfect metamorphosis, as the Grasshoppers and the Locusts. They are produced from the eggs without wings, but have them formed gradually while they are in a state of activity.

Questions.—What are the three changes which most Insects pass through in their metamorphosis? Explain the terms larva, pupa, and imago. What are the different names applied to larvæ? What is said of the terms chrysalis and aurelia? What is said of the contrast between the larva and imago states? What is said of the growth of the larva? What of the growth of the imago? What of the casting of the skin of the larva? What is said of the various modes of passing the pupa state? Describe the manner in which the Silkworm makes its cocoon. Describe the mode of its exit. What is said of the thread of which the cocoon is made? What are the native countries of the Silkworm? What is said of the silk manufacture? When pupæ are to be out doors all winter, what provisions are adopted to guard against the cold? Describe the Cecropia Moth Describe its caterpillar. Describe the construction of its cocoon What is said of its exit? What is said of the manufacture of silk from the cocoons of this and other large Moths? What is said of Grasshoppers and Locusts?

CHAPTER XXIV.

COLEOPTERA, OR SHEATH-WINGED INSECTS.

419. I now pass to the consideration of the different orders of Insects, noticing particularly some of each order. They are arranged in orders according to the character of the wings. They are chiefly the following: 1. Coleoptera (κολεος, *koleos*, a sheath, and πτερον, *pteron*, a wing), Sheath-winged. This is the order of Beetles. 2. Orthoptera (ὀρθὸς, *orthos*, straight, and πτερον), Straight-winged. This includes the Grasshoppers, Locusts, etc. 3. Neuroptera (νεῦρον, *neuron*, nerve, and πτερον), Nerve-winged. 4. Hymenoptera (ὑμεν, *humen*, a membrane, and πτερον), Membrane-winged, including the Bees, Wasps, etc. 5. Lepidoptera (λεπις, *lepis*, a scale, and πτερον), Scale-winged. The Butterflies and Moths. 6. Hemiptera (ἡμισυς, *hemisus*, half, and πτερον), Half-winged, including Bugs, Cicadæ, etc. 7. Diptera (δις, *dis*, twice, and πτερον), Two-winged. Flies, Musquitoes, etc. 8. Aphaniptera (ἀφανὴς, *aphanes*, not manifest, and πτερον). The Fleas belong to this order. 9. Aptera, Wingless. The prefix *a* in this case is privative or negative. The common louse, sugar-lice, spring-tails, etc., belong to this order. There are some other orders, which are small, however, and of little importance.

420. The order of Coleoptera, or Sheath-winged Insects, is the most numerous of all the orders. "It is probable," says Carpenter, "that from thirty to forty thousand species of Beetles alone now exist in the cabinets of collectors; and we may safely affirm that at least as many more remain to be discovered." They are of various size, some being very small, and others among the largest of Insects. There are some that are five

inches in length. The Beetles, when not flying, appear to have no wings. The elytra, § 398, are horny, and fit closely together on the back. These are sometimes very beautiful, even splendid, having various colors, golden, green, blue, etc. I have seen them in some cases with depressions in them spotted with gold, exactly as if real gold-leaf were inserted by powerful pressure. The wings, which are commonly twice as large as the elytra in length as well as in breadth, are folded up under these covers very curiously when the insect is not on the wing. There are only a few exceptions in the whole order to this mode of arrangement.

421. The metamorphosis in this order is complete. The larvæ are worm-like, having soft bodies, but they commonly have horny heads. Those which lead a retired, still life, as those which are in nuts, have no legs, for they need none. Those larvæ which are carnivorous have the strongest legs. In some of the herbivorous species, besides the true legs, there are fleshy tubercles, which are called pro-legs. Previously to entering the pupa state the larva often forms a case for itself of bits of earth, or of chips, which it unites together by silken threads or with a gluey substance. The pupæ of some Beetles are inactive for years.

422. Beetles, as suggested by Professor Jaeger, are of three kinds: 1. Carnivorous Beetles: they devour living insects, and are the beasts of prey of the insect world. 2. Scavenger Beetles, which live on putrid matter, carrion, and decaying vegetable substances. 3. Herbivorous Beetles, which feed on living plants and fruits. The first two kinds are of great use to man, but the last are injurious. I will notice a few of each kind.

423. Of the Carnivorous Beetles, the Lady-birds, so called, are known to almost every one. They look like little colored and spotted turtles. The larvæ of these Beetles are of great service to man, for they prey upon the plant-lice which are so destructive to many plants,

being more effective in this respect than the Beetles themselves. They are about half an inch long, of a bluish color with four or six yellow spots, and are seen creeping along branches and leaves in search of the plant-lice. After living in the larva state for a fortnight,

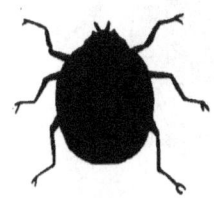

Fig. 190.—Northern Lady-bird.

they fasten themselves on some leaf, cast off their skins, spin a cocoon, and, after a fortnight's sleep, issue as Lady-birds. One of the species, the Northern Lady-bird, is seen in Fig. 190. It is found chiefly on the pumpkin vine, where, in the company of their larvæ, they feed on the Plant-lice and the larvæ of the Squash Bug.

424. The Tiger Beetles, of which two species are

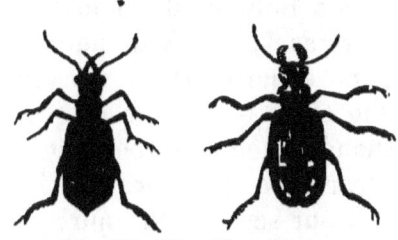

Figs. 191, 192.—Tiger Beetles

represented in Figs. 191 and 192, are thus named both on account of their variegated colors and their rapacity. They feed on Caterpillars, Flies, and other Beetles, and will even devour each other when shut up together. Their larvæ or grubs are as rapacious as they are themselves. They live in holes which they dig in the ground. When they are hungry they come to the mouth of their holes, and there keep watch, seizing the first insect that passes over the hole. Though these grubs are soft and white, they have powerful and well-armed jaws, with which they gratify their rapacity.

425. In Figs. 193 and 194 (p. 248) you have two species of Caterpillar-hunters, so called from their habits. They are very handsome Beetles. The green Caterpillar-hunter, Fig. 193, is a great devourer of the Canker-worm. These Beetles run about in the grass after the worms, and go up the trunks of the trees to capture them as they come down.

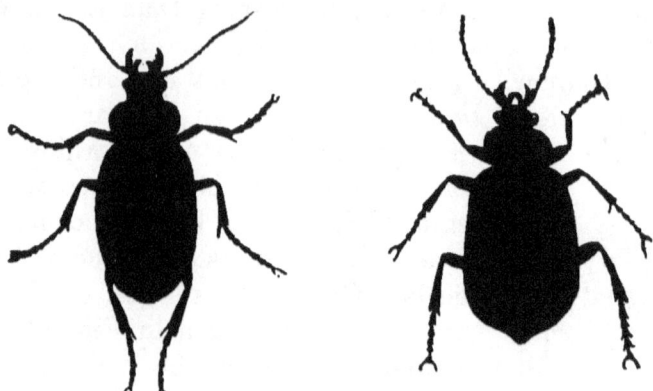

Figs. 193 and 194.—Caterpillar-hunters.

426. The Scavenger Beetles, forming the second division, have very fine coverings, and their feet are fitted for digging. Though they are not only in the midst of filth, but live on it, they are remarkably clean, and are generally of a bright color, and some of them are very beautiful. These Beetles, and their grubs, are of great service as scavengers. Although each one does but little, the multitude of them clear up a great deal of filth, which would otherwise offend our senses and injure the health. Those Beetles of this class which are of very large size, sometimes five inches in length, are found in the tropical regions of America, Asia, and Africa.

427. To this class belong the common Tumble-bugs, or Pellet Beetles. These exhibit great industry in rolling balls of manure and earth mingled together. In every one of these is deposited an egg. These busy animals dig holes two or three feet deep, and into these they roll the balls. While they thus provide for their progeny, they are at the same time useful in distributing manure among the roots of plants where it is wanted.

428. To this group also belong the Carrion Beetles, of which there are many species, some of them very beautiful. They are exceedingly busy wherever there is the dead body of any animal, devouring it and depos

iting in it and upon it their eggs. The Crusader Carrion Beetle, Fig. 195, is thus named by Jaeger from a black spot on the back of its yellow thorax, which resembles somewhat the figure of a cross which the Crusaders wore on their coats. The wing-covers, or elytra, are brown, and the head and legs are black. These Beetles are seen in immense multitudes in some carrion. The habits of the Big Gravedigger are very curious. It gathers in great numbers round a dead frog, or mouse, or bird, etc. The Beetles first examine the spot where the dead body lies; if it be stony, they select a proper place, and, by their combined efforts, remove the body there. They now proceed to dig the earth away from under it with their fore feet, and do not leave it till they have sunken it about a foot in the ground.

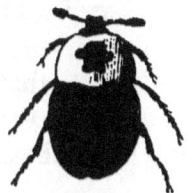

Fig. 195.—The Crusader Carrion Beetle.

429. There is a very small Beetle of this class which is a great destroyer of the collections of the naturalist, eating the skins of stuffed animals, and the internal parts of insects. It is hence called the Cabinet Beetle. It is difficult often to keep a cabinet free from them, for their larvæ will eat through the hardest boards.

430. Among the Scavenger Beetles are some wood-eating insects. These are of great service in tropical countries, where large trees are prostrated in great numbers by hurricanes and tempests. It would take a long time for the natural process of decay to remove them; but these insects reduce them to dust in a short time, and this dust, becoming incorporated with the earth, fertilizes it. The common Horn Bug or Stag Beetle belongs to this group. The Stag Beetle of Europe is twice as large as that of this country. It is the grubs that live on wood, and not the Beetles themselves. The grubs of some of the wood-eating Beetles are some years in arriving at the mature state in which they are ready to change into Beetles.

431. The Herbivorous Beetles live on vegetable food both in their larva and imago state. Some eat fruits, some grain, some leaves, and some even wood. Science has been of great service in pointing out the Insects that inflict these various injuries, and also in indicating the means of prevention by discovering the habits of these animals.

432. To this class belong the Spring Beetles, sometimes called Skippers or Snapping Bugs. They are so constructed that when they are laid on the back they can throw themselves upward, and coming down alight on their feet. This performance, which is a great amusement to children, is done by a spring which the animal has in its body for this purpose. The largest and handsomest of these Beetles in the United States is the Velvet-spotted Spring Beetle. Another species of the same genus is the Lightning Spring Beetle, Fig. 196. This Insect, which is nearly an inch and a half long, has two yellow cone-like projections on the sides of the thorax, which emit light, and appear, while the animal is alive, like two shining emeralds. It also emits light from the under surface of the segments of the abdomen. In Cuba ladies fasten these Beetles in their hair as ornaments at evening parties. The light of our common Fireflies is emitted from two or three segments of the abdomen, as you may see by catching one, and holding it in your hand turned over on its back.

Fig. 196.—The Lightning Spring Beetle.

433. The Capricorn Beetles are so named from the resemblance of their long antennæ to the horns of the Mountain Goat. These Beetles are very beautiful, although their grubs are ugly. The Painted Capricorn appears with us in the autumn, and may be seen in the

flowers of the golden-rod. Its body looks like black velvet. Its head and chest are crossed with yellow lines, and its elytra have lines of the same color variously arranged. The female deposits her eggs in the crevices of the bark of locust-trees, and the grubs hatched from them eat the wood and the pith, making winding passages in doing this, and, of course, proving destructive to many of these trees. There is in the southern parts of our country a Beetle of this family three and a half inches long, called the Stag Beetle Capricorn, because its formidable jaws, an inch in length, are like those of the Stag Beetles. In South America is found a splendid Beetle, called the Long-armed Capricorn, its fore legs being five inches in length. It is of a dark olive-green color, with markings of red, yellow, and white, resembling hieroglyphics.

434. The Spanish Fly, which is used in making the common blistering plaster, is an herbivorous Beetle. It has a brilliant green metallic hue. It is the powder of the dried Beetles that makes the basis of the blistering salve—another example of animal chemistry both wonderful and mysterious, § 170.

435. The Curculios or Weevils are a family of herbivorous Beetles that do great injury to fruits and grains. The perfect insect deposits its eggs in them, and the grubs or maggots live on the substance in which they are hatched. Thus a little hairy gray Beetle deposits its eggs in the young and tender peapod, and the larvæ hatched from them eat portions of the peas as they grow. Multitudes of these larvæ are boiled in the peas prepared for the table. So also in almost every seed-pea there is either a Beetle or an opening from which one has come out. In the same way the maggot found in the chestnut comes from an egg deposited in it by a Beetle in an early stage of the fruit. So also in the apple and other fruits.

436. These Weevils, or Snout Beetles, have an appa

ratus for boring holes in the grains and fruits, as you see in the Palm Weevil, Fig. 197. "Its larvæ," says Jaeger, "are known in the tropics of America under the name of Palm-worms, and they live in large numbers in the trunks of several palm-trees, but principally in the cabbage-palm, which grows in abundance in the mountainous parts of St. Domingo. When fully grown they are about three inches long and one inch in circumference, of a dirty yellow color, with a black head, looking like a piece of fat enveloped in a transparent skin. These disgusting-looking animals are roasted upon a wooden spit, or broiled and eaten with dry and pulverized bread, seasoned with salt and pepper, and considered by many epicures as the *ne plus ultra* of delicacies."

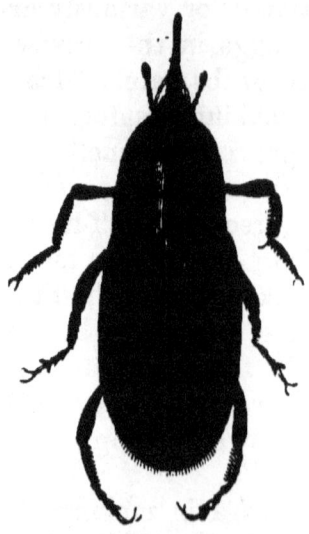

Fig. 197.—Palm Weevil.

437. The Leaf-eaters, which live mostly on leaves and flowers, are very small Beetles, very richly colored. Among the most brilliant is the Gilded Dandy, Fig. 198, found abundantly on the dog-bane in July and August. The larvæ or grubs of the Leaf-eaters have six legs, as they must crawl about in getting their food, instead of remaining in one spot as the fruit-eating grubs of the Weevils do.

Fig. 198.—Gilded Dandy.

Questions.—Name the orders of Insects, and give the chief characteristics of each. What is the extent of the order Coleoptera? What is said of the size of the insects belonging to it? What is said of the elytra? What is said of the larvæ of Beetles? What are the three kinds of Beetles? What is said of the Lady-bird? What of the Tiger Beetles? What of the Caterpillar-hunters? What of the Scavenger Beetles? Of the Pellet Beetles? Of the Carrion Beetles?

Of the Crusader Carrion Beetle? Of the Big Gravedigger? Of the Cabinet Beetle? Of the Wood-eating Beetles? What is said of the herbivorous Beetles? Of the Spring Beetles? Of the Lightning Spring Beetle? What gives the name to the Capricorn Beetles? What is said of the Painted Capricorn? Of the Stag Beetle Capricorn? Of the Long-armed Capricorn? Of the Spanish Fly? What is said of the Curculios? What of the Palm Weevil? What of the Leaf-eaters?

CHAPTER XXV.

STRAIGHT-WINGED INSECTS.

438. The second order is that of the Orthoptera, or Straight-winged Insects. Their wings, when not in use, are folded lengthwise like a fan, and are extended straight along the top or the sides of the back. These are covered by a pair of thicker wings, or, rather, wing-shaped members, which in the Grasshoppers and the Locusts are long and narrow, and are joined together on the back, making two slopes like the roof of a house. These wing-covers are intermediate between the stiff, horny elytra of the Beetles and the membranous wings of some other insects.

439. The insects of this order do not go through with a complete metamorphosis. They do not pass at all into the torpid pupa state, but are active during the whole period of their existence. At first they are destitute of wings; but they become winged as they grow, casting off their skins about six times during the process. They are divided into four families: 1. The Cursoria, or Runners. 2. The Raptoria, or Graspers. 3. The Ambulatoria, or Walkers. 4. The Saltatoria, or Jumpers.

440. The family of Cursoria includes the Cockroaches and the Earwigs. There are with us two kinds of Cockroaches—the native ones, found under stones in the field, and those which have, like the Rats, been introduced from other countries, and live in our houses. These vo

racious animals, troublesome as they are here, are vastly more so in some other countries. It is said that some houses in St. Petersburg became so infested with them that no one could live in them, and they were burned down to destroy these insects.

441. Earwigs are little insects having a pair of nippers, shutting like scissors, at the hinder end of the body. They eat both fruit and flowers, disfiguring the latter with holes. They are very timid, running for some crevice whenever disturbed, and thinking that they are safe if they put their heads under cover, and thus get out of sight of danger. They are apt, when frightened, to plunge down into the bottom of a flower, if they happen to be on one, leaving, however, their curious forked tails standing up among the stamens. Their name is not an appropriate one, for they have really never been known to enter the human ear. These insects are very different from the animal so often called by this name in this country, which is really not an insect.

442. Among the Raptoria is that singular insect the Mantis Religiosa, or Praying Mantis, Fig. 199. It is so

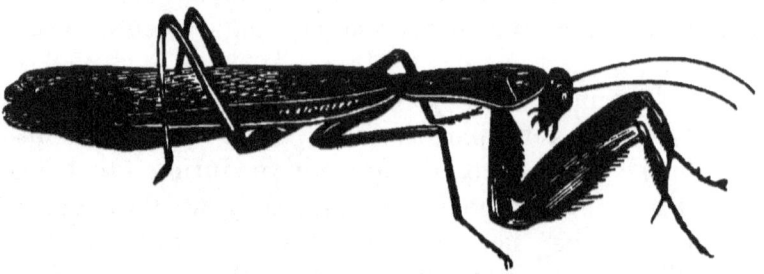

Fig. 199.—Mantis Religiosa.

called from the attitude which it assumes when it is watching for its prey. The front of its thorax is raised, and the two fore legs are held up together, like a pair of arms, ready to seize any insect that may come within its reach. These insects are extremely voracious. If two are kept together without food, they fight until one is killed, and the victor devours his adversary. Fights

STRAIGHT-WINGED INSECTS. 255

between these insects are among the sports of the Chinese, the pleasure being the same as that which is derived from cock-fights and bull-fights.

443. The family of the Ambulatoria is a very small one, including those very singular animals, Walking-sticks, Walking-leaves, etc. They lead a sluggish life among the branches of shrubs, living on the young shoots. Their color and shape being so much like those of things around them, enable them commonly to escape observation. Some of them, as the Walking-stick, Fig. 200, have no wings, and look like dead twigs, the legs

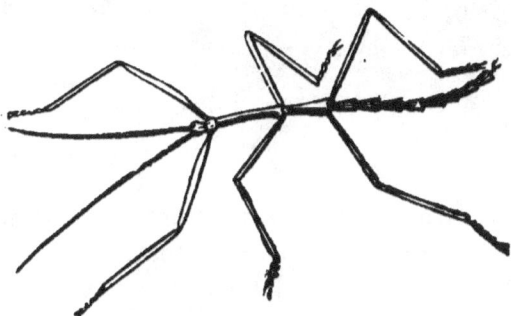

Fig. 200.—Walking-stick.

appearing like little branches. There are found of this insect twenty species in South America, three in North America, three in Europe, forty in Asia, twenty-seven in Australia, and two in Africa. The Leaf Insect, Fig. 201, is of the same family. It is found in South America

Fig. 201.—The Leaf Insect.

It resembles a leaf both in shape and color, and the wings have even the veinings of a leaf.

444. The family of Saltatoria, or Jumpers, is a very extensive one. It comprises the Crickets, the Grasshoppers, and the Locusts. The Crickets are so well known to you that I need not describe them. They are mostly inhabitants of the ground, in which many of them burrow. One species, the Mole Cricket, Figure 202, is so named because its anterior extremities, and its general habits also, are similar to those of the Mole. It is a great digger. The female forms, in connection with its burrow, a smooth, round cell, which, with the passage leading to it, resembles a bottle with a long bent neck. Here it deposits from two to four hundred eggs.

Fig. 202.—The Mole Cricket.

The Tree Cricket, Fig. 203, is a very delicate insect. Its color is pale ivory; its antennæ and legs are very long, and its wing-covers are thin, and are prettily ornamented with three oblique raised lines. Its familiar shrill sound is produced only by the male Cricket, by raising up the wing-covers and rubbing them together. These differ decidedly from the other members of the Cricket tribe in living wholly on trees. The female deposits her eggs in the autumn, in incisions which she makes in the branches, and they are hatched in the following summer, the young Crickets obtaining their perfect state with us in August.

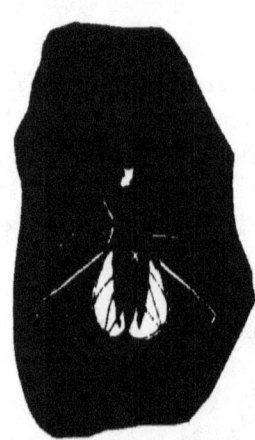

Fig. 203.—Tree Cricket.

445. The Grasshoppers differ from the Crickets in having the wing-covers, which in the latter lie horizontally flat, so arranged as to make two slopes, like the roof of a house. Of the many species I will notice but one, the well-known Katydid of this country. It is about one and a half inches long, and its expanded wings measure together three inches. The whole insect is green, the wings being pale green, and the wing-covers a dark green. The wings are gauze-like, and are exceedingly delicate. The male, as seen in Fig. 204, has, at the base

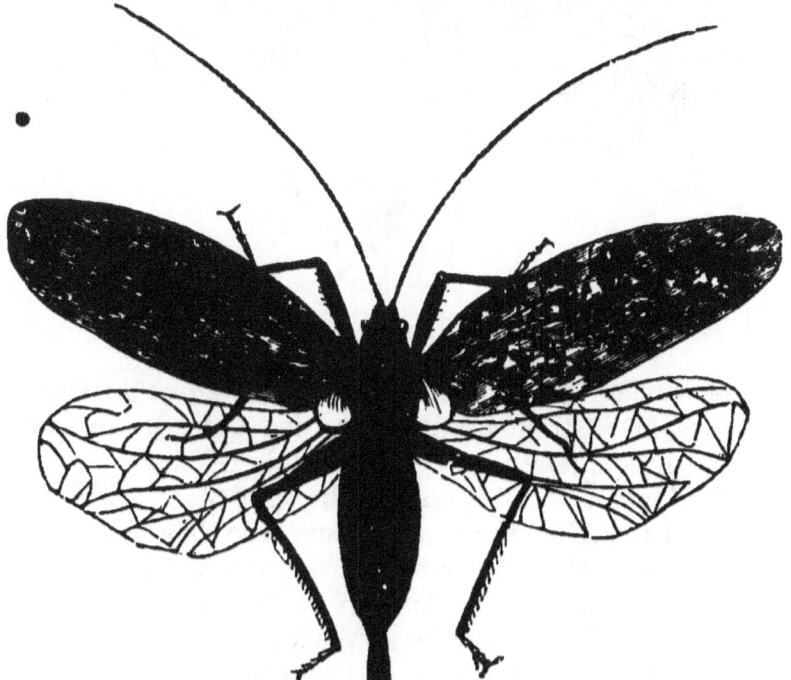

Fig. 204.—Male Katydid.

or root of each wing-cover, a stout horny ridge surrounding a stiff, thin membrane, making two drum-heads. It is by the rubbing of these together that the peculiar sound of this insect is produced. The female Katydid has no such apparatus, and therefore is perfectly still. It has at the end of its body, as seen in Fig. 205 (p. 258),

Fig. 205.—Female Katydid.

a sword-like instrument: this is called an *ovipositor*—that is, an instrument for depositing the eggs in their right place. With it the insect pierces holes in the ground in the autumn, placing the eggs there, which are hatched the following year. Some time elapses between the birth of the Katydids and the attainment of their growth, and the full production of their wings, which is necessary to the production of the loud sound with which they greet our ears at night in the latter part of summer.

446. The Locusts, one species of which you see represented in Fig. 206, have, like the Grasshoppers, the roof-like arrangement of the wing-covers, but they have not that peculiar apparatus for the production of sound which the Grasshoppers have. They have also shorter antennæ and stouter legs. They are insects of greater power. In some parts of the world they are the most extensive-

Fig. 206.—Locust.

ly destructive of all insects. Some species are occasionally quite destructive in some parts of this country. But it is in Asia and Africa that they appear in such immense armies, leaving not a vestige of vegetation in their track, eating the corn and the grass down to the roots, and stripping the trees of their leaves. Mr. Cumming, in describing the flight of an army of these insects, says, "I stood looking at them until the air was darkened with their masses, while the plain on which we stood became densely covered with them. Far as my eye could reach, east, west, north, and south, they stretched in one unbroken cloud, and more than an hour elapsed before their devastating legions had swept by." These insects sometimes make incursions into Europe. One of these is described by Professor Jaeger, who was an eyewitness of it as he was traveling in Russia in 1825 across its desert prairies. The carriage-wheels moved through Locusts piled up to the height of two feet. This state of things existed over a wide extent of country. The insects were now wingless; but the inhabitants of the fertile regions north feared that, as soon as their wings were grown, they would come north and devour every green thing. Before this vast insect army could do this, the Emperor

Alexander sent an army of thirty thousand men against it. "The soldiers," says Jaeger, "forming a line of several hundred miles, and advancing toward the south, attacked them, not with sword and gun, but with more ancient implements—with shovels. They collected them, as far as possible, in sacks, and burned them." Notwithstanding this war upon them, the vegetation was destroyed by them to a great extent.

447. The ravages of the Locust are often adverted to in the Bible, and the descriptions there given correspond with those of modern travelers. They are spoken of as a "great army," and it is said that "the land before them is as the Garden of Eden, and behind them a desolate wilderness"—a result often witnessed at the present day. The manner in which this insect army makes its invasion is most graphically described in the second chapter of Joel.

448. Some species of Locusts are eaten now in the East as they were in the time of John the Baptist. Mr. Cumming, a traveler in South Africa, thus speaks of them as food. "Locusts afford fattening and wholesome food to man, birds, and all sorts of beasts; cows and horses, lions, jackals, hyænas, antelopes, elephants, etc., devour them. We met a party of Batlapis carrying heavy burdens of them on their backs. Our dogs made a fine feast on them. The cold, frosty night had rendered them unable to take wing until the sun should restore their powers. As it was difficult to obtain sufficient food for my dogs, I and Isaac took a large blanket, which we spread under a bush whose branches were bent to the ground with the mass of Locusts which covered it, and, having shaken the branches, in an instant I had more Locusts than I could carry on my back; these we roasted for ourselves and our dogs."

Questions.—What is the arrangement of the wings of the Orthoptera? What is said of the metamorphosis of this order? What are its four families? What does the family Cursoria include? What is said of the Cockroaches? What of the Earwigs? What of the Praying Mantis? What of the Walking-stick? Of the Leaf Insect? What does the family Saltatoria include? What is said of the Mole Cricket? Of the Tree Cricket? How do the Grasshoppers differ from the Crickets? What is said of the Katydid? In what way is its sound produced? How do the Locusts differ from the Grasshoppers? In what countries are they at times exceedingly numerous? Describe their appearance in Russia in 1825, and the means taken to destroy them. What is said of their ravages? Give the description from the Prophet Joel of the invasion of an army of Locusts What is said of these insects as food?

CHAPTER XXVI.

NET-WINGED INSECTS.

449. THE insects of the order Neuroptera, or Net-winged Insects, have, like the Coleoptera and Orthoptera, a mouth fitted for mastication, but differ from them in their wings. They have no wing-covers, but there are commonly four thin and transparent wings, with the veins forming a delicate net-work, as seen in Fig. 207 (p. 262). The posterior wings are ordinarily as large as the anterior, but in some species they are quite small, and in some few entirely absent. The body is long, slender, and soft. These insects are of intermediate size, none being either very large or very small. There are about a thousand species. The metamorphosis is not alike in all. In some it is complete, the larva having a form very different from the imago or perfect insect, while in others there is little difference except in the absence of wings in the larva and their presence in the imago, as in the Grasshoppers and Locusts. By these differences the order is naturally divided into two groups, in the first of which the insect is active during its pupa state, while in the other it is torpid during this state, except just before its last metamorphosis. Of the first group there are five families, the Dragon-flies, Day-flies, Stone-flies, White Ants, and Book-

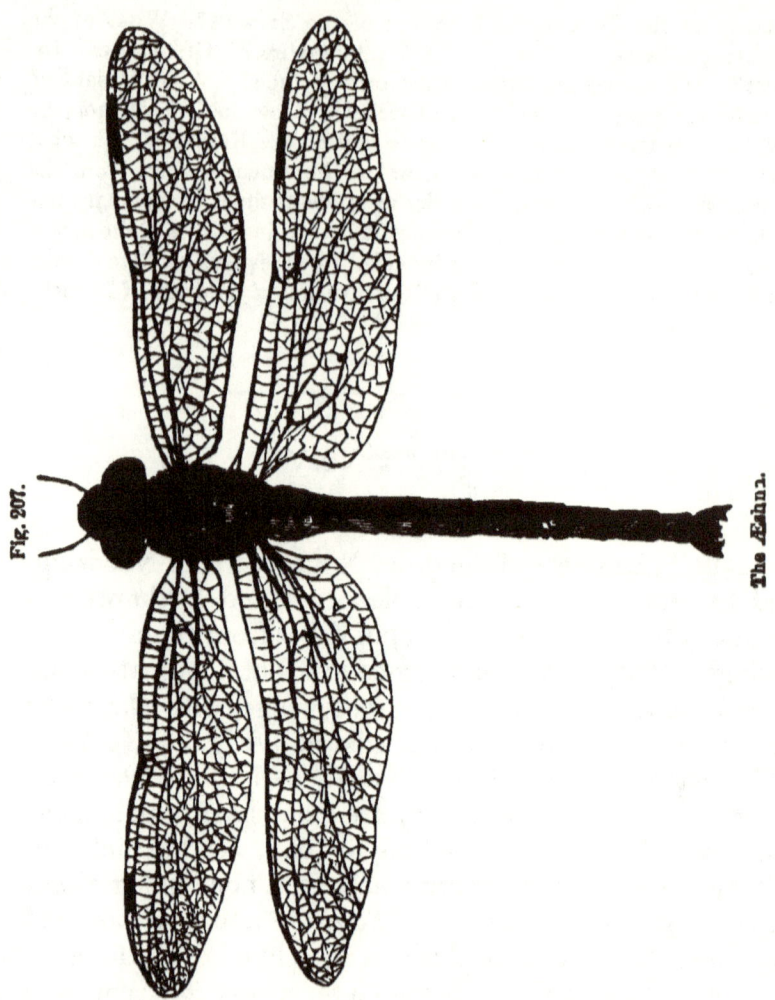

Fig. 207. The Æshna.

lice. In the first three of these families, the insect in its larva and pupa states inhabits the water, respiring like a fish, having peculiar organs for this purpose at the sides of the abdomen or at its end, while in the form of its body it is quite like the insect in its perfect state.

450. Of the Dragon-flies, the most conspicuous and best-known family, there are about two hundred ascertained species. These are the Swallows of the Insect

NET-WINGED INSECTS. 263

tribe; for they catch their prey, which consists of Flies, Musquitoes, Butterflies, etc., on the wing. They do this, however, with their claws, and not, like the Swallows, with their mouths. That they may readily see their prey as they fly about so swiftly in search of it, they have very large, compound eyes, as you see in Fig. 207, one of our common Dragon-flies, or Darning-needles, as they are often called. These formidable-looking insects are entirely harmless, never biting or stinging when we catch them. They are of great service to us in destroying the Musquitoes, of which they devour a great number. Some species are beautifully variegated in color.

451. The eggs of these insects are deposited on the leaves of aquatic plants. The larvæ live wholly in the water. They have some very singular peculiarities. They have a kind of mask with which they can cover up their mandibles and most of the head. But this mask can be unfolded and extended, and, having on its end a pair of claws, it is used as an instrument for seizing their prey, as represented at A in Fig. 208. At B the insect is seen with the mask folded up. You see here, also, water issuing from the end of the larva's body. It is in this way that it propels itself through the water, just as a rocket

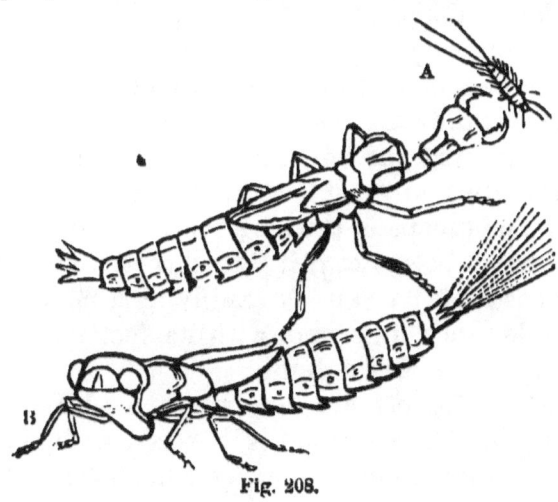

Fig. 208.

rises by the stream of fire at its end. Another purpose is accomplished, also, by this operation. The breathing apparatus is in this quarter of the body, the air in the water being there introduced to the blood of the insect, just as it is introduced to the blood of the fish in its gills. It spends nearly a year in the water, and then comes its metamorphosis. This, Jaeger says, "may be observed almost daily from the month of April until October, but occurs principally in the months of May and June. But this transformation does not take place in the water, but out of it; and when ready for their metamorphosis, the larvæ climb up the stem of some water-plant, and in about two hours after are capable of raising themselves up by their wings and flying away in the air. This whole operation may be witnessed by putting the grubs into a pail of water, and placing in it some sticks or branches upon which they may creep up and prepare themselves for their aerial journeys."

452. The Ephemeridæ, or Day-flies, are so called from their short existence in the imago state, which, like that of some flowers, is limited to a single day. In their larva state, however, they have a long life of two or even three years. During this time they are inhabitants of the water, having leaf-like appendages on their sides as their gills or respiratory apparatus. When they are about to change to the imago state, wings are formed, but are kept folded up till they are ready to leave the water. While these are forming the insect is said to be in the pupa or chrysalis state, and yet it is as active now as when it was a larva. The escape of the insect into the air is so quickly done, that it seems as if it flew directly out of the water. It casts off its skin as readily as a man puts off a coat, unfolds its wings, and, with its feet resting on its cast-off skin, it takes its flight.

453. These insects are sometimes produced in such multitudes that the ground is covered with their dead bodies, and they are carted away as manure. Professor

Jaeger saw great numbers of them once in the Raritan River, near Trenton; but the greatest display of them that he ever witnessed was in the River Neva, in Russia. "The light of the sun," he says, "was intercepted as in a thick fog, so much so that nothing could be distinguished at the distance of a few yards. The atmosphere had something the appearance it presents in a violent snow-storm, and thousands of Day-flies fell into our boat and all over our persons; while the fishes in the water, the birds in the air, and the domestic fowls upon the shore, were every where feasting upon them." He farther says, "In the evening these flies are strongly attracted toward a light, perhaps more so than any other nocturnal insect, and it is very amusing to see the crowds of them that fly through an open window and dance around the light, making a variety of turns, and circles, and waltzes. They fly so close together, and glisten with such splendor, that the observer sees a ribbon of gold continually revolving around the light, or imagines a celestial globe of living circles revolving in every direction, while the light represents the central sun."

454. The Termites, or White Ants, are the only family of the order Neuroptera that live in communities with a regular social organization. They are, with some few exceptions, confined to tropical climates. Next to the Locusts, they are the most destructive of insects, as not only food, but clothing, trees, fences, and even houses, are devoured by them. One species has lately done great damage in France. While they are thus destructive, they are, considering their size, the greatest of all builders, going far beyond man in this respect. Their habitations are some ten or twelve feet high, having much the shape of a sugar-loaf. They are built of clay, which these insects in some way render as hard as some kinds of stone. There are various apartments and winding passages in this dwelling, and there are passages dug in different directions under ground, all lined with the hardened clay.

These galleries are sometimes carried under houses, which the Ants enter, and, eating out all of the inside of the timbers, leave them only as mere shells. Sometimes there are many of these curious structures in the same neighborhood. Dr. Adamson says that, in some parts near Senegal, there are so many of them near together that they appear like native villages.

455. The community in one of these habitations is immense in number, consisting of laborers and soldiers under a king and queen. These last are the only ones that come to the imago or perfect state. The laborers seem to be larvæ stopped in their development, so that they never acquire wings. The soldiers, on the other hand, are pupæ. The queen lays all the eggs, to the number, it is estimated, of forty or fifty millions in a year. This she does in a royal chamber set apart for this purpose. The laborers take the eggs as fast as she lays them, carry them away to the nurseries, where they are hatched, and take care of the young. They also do all the building and repairing, gather all the stores, and perform all the labor of any kind that is needed. The soldiers, on the contrary, do no work, but stand guard, and defend the community, in which they show great bravery and energy, appearing boldly upon the outposts when any enemy appears, while all the laborers retire within. The royal chamber is near the centre of the hillock, and is surrounded by apartments which are occupied by what may be called the body-guard of the queen, some of the soldiers, and by her immediate attendants, some of the laborers. She can never leave her chamber, for no opening from it is large enough for the passage of her body, which is enormously enlarged for the production of its multitudes of eggs. The minutiæ of the arrangement of the nurseries and the various apartments, and of the economy of this wonderful community, are very interesting, but can not be entered upon here.

456. The Book-lice form a small family nearly allied

to the Termites. They usually live in damp, dark places, and under bark. One species, destitute of wings, is often found in old books, and in collections of dried plants, insects, etc.

457. In the second division of the Neuroptera the metamorphosis is more complete, the pupa being inactive. The most singular of these insects is the Ant-lion. The wingless larva has a curious contrivance for securing its prey, which consists of ants and other insects. It digs a funnel-shaped pit in sand, about thirty inches in diameter and twenty inches deep. This is an immense work for so small an insect. It accomplishes it in this way. It first traces the circle which is to be the outer edge of the pit. Then, placing itself within this line, it, with one of its legs for a spade, places some sand in a heap on its head, which with a quick jerk it throws beyond the circle to the extent of some inches. It does this around the whole circle; then turning, goes round again, and so on until the whole pit is dug. It now conceals itself at the bottom, and watches for some insect to tumble down into the pit. If the insect does not fall to the bottom, and endeavors to escape, the Ant-lion, with its head and mandibles, throws over it a quantity of sand, and thus overwhelms its victim. Of course, such a struggle disturbs the evenness of the pit, and the breaches are immediately repaired, so as to be in readiness for other prey.

458. There is a family belonging to this section, called Hemerobiidæ, remarkable for the brilliancy of their eyes, and the delicacy and varied color of their wings, but especially for the singular manner in which they dispose of their eggs. They deposit them usually upon plants, at the end of long and exceedingly delicate footstalks, the base of which is firmly attached to the leaf. These footstalks are composed of a white viscid matter, discharged at the time of laying the egg, and speedily hardening in the air. As these eggs are laid in clusters, the appearance is that of small clusters of fungi. I saw some once

upon a pane of glass, and here the breadth and firmness of the attachment of the base of each footstalk were very manifest, seen through a small microscope.

459. There is an aberrant family of this section called Caddice-flies, remarkable for the covering of hair with which both their bodies and wings are beset. The habits of their larvæ are very interesting. They are aquatic, and live in cylindrical cases open at each end. To these cases they attach various substances, such as bits of wood, weeds, pebbles, shells, etc. In Fig. 209 are represented several of these tubular houses with various things attached to them. The different species, of which there are many, seem to have their individual preferences in relation to the substances which they employ; but they readily disregard these preferences when there is a lack of those materials which they usually prefer. They never willingly leave their cases, but only thrust the head and a portion of the body out in search of their food. When about to pass into the torpid pupa state, they fasten their tubular houses to something in the water, and then close the two ends with a kind of silken grating which allows the water to pass freely through it. When they are to assume the imago form, they make a hole in the grating with a pair of hooked jaws which they now have. They are now good swimmers, using chiefly their hind legs for this purpose. Coming to the surface of the water, and perhaps climbing up some plant, the skin of the swimmer gapes open, and out flies an insect about double the size of that

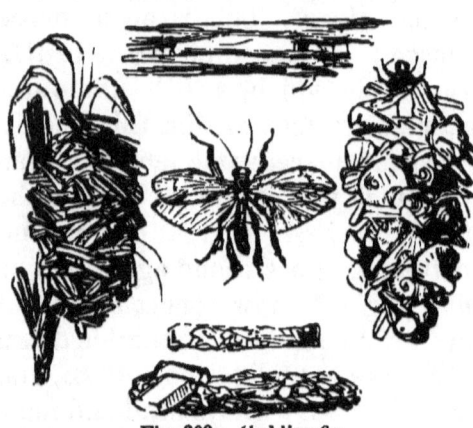

Fig. 209.—Caddice-fly.

which you see in the centre of Fig. 209. Fishes are very fond of the larvæ of the Caddice, and hence the necessity of such a covering as they make for themselves. For this reason, also, they are often used as a bait by the angler.

Questions.—In what are the Neuroptera like the Coleoptera and the Orthoptera? What is said of their wings? How are they divided into two groups? What are the families of the first group? What is there peculiar in three of these families? What is said of the Dragon-flies? What of their eggs? What of their larvæ? What of their metamorphosis? What is said of the Day-flies? Give the narration of Jaeger in regard to them. What is said of the ravages of the Termites? What of their habitations? What are the different classes in their communities? What is said of the laborers? Of the soldiers? Of the queen and her cell? What is said of the Book-lice? What is the characteristic of the second group of the Neuroptera? What is said of the Ant-lion? What is said of a family of insects that deposit their eggs on stalks? Give the account of the larvæ of the Caddice-flies. Describe their pupa state and their metamorphosis.

CHAPTER XXVII.

MEMBRANE-WINGED INSECTS.

460. THE wings of the insects of the order Hymenoptera are membranous, like those of the Neuroptera, but differ from them in not having a fine net-work of veins or nerves.* In some of the very small species there are almost no nerves. The name membrane-winged is therefore more appropriate than vein-winged, which is sometimes given to them. The anterior wings of the Hymenoptera are usually much larger than the posterior, and during flight the wings of each side are fastened together by minute hooks on the posterior wing, which take hold of the rear margin of the anterior one. The females

* These two terms, meaning the same thing, must not be confounded with the same terms used in their ordinary sense. In insects they are applied to the frame-work of the wings.

have a peculiar prolongation of the last segment, which in one division of the order is an ovipositor, and in the other is a sting. The Hymenoptera are farther distinguished by a remarkable development of the instinctive faculties, especially those which have a complicated social organization of their communities, as the Bees.

461. The metamorphosis in this order is complete, the pupæ being quite inactive, and the larvæ are more imperfect than in any other order. In most of the species they have no feet, and resemble worms. The larvæ of some of them, however, are like caterpillars, and have eighteen or even twenty feet. Jaeger says of them that they live "in clean places, such as cells artificially built of wax, pieces of wood, leaves, or mortar; or they dwell in wood, in holes under ground, in gall-apples, or oak-balls, and many live in caterpillars; but none inhabit carrion, dunghills, or other putrid and filthy places."

462. None of the Hymenoptera are very large, and some are exceedingly small. In numbers this order is inferior only to the Coleoptera, and it has been estimated to contain one fourth of the whole insect world. Though the Hymenoptera are the most numerous and largest in tropical countries, they are widely distributed in almost every part of the earth. They are mostly great workers, and none are nocturnal, but all do their work in the day. Some of them are very useful to man, the Bees supplying him with honey and wax, and the Gall-insects with a material valuable for making ink, and especially for coloring.

463. We divide the order into two groups: 1. The Terebrantia, or Borers, whose females have ovipositors; 2. The Aculeata, or Stingers, in which the females have a sting, or piercer, connected with a reservoir of poison. I will notice but a few families in each group.

464. The family of Gall-flies is one of the most prominent among those of the first group. These insects, with their ovipositors, make slits in various parts of

plants and trees, depositing therein their eggs. They moisten these cuts also with an irritating fluid, which causes the growth of the tumors called *galls*. When the eggs are hatched, the larvæ live on the interior of these tumors, just as the larvæ of the Nut-weevil, § 435, live on the nuts in which they are hatched. It is remarkable that the same tree should produce on its different parts galls of various forms and degrees of hardness, according to the species by which the eggs are deposited. The hardest gall is the common gall-nut of commerce, so much used in making ink, and in the process of dying black. This is the product of an oak growing in the Levant. It has been found that the famous "Apples of Sodom" are galls of a different consistence on the same oak, occasioned by another species of Cynips, or Gall-fly. While the oak apples, so familiar to us, appear on the twigs of the oak, there are also different kinds of galls produced on the leaves, the catkins or pendent flowers, and even on the root. Those on the root are large and woody, and eleven hundred larvæ have been found in a single one of them. While the oak seems to be a great favorite of the Gall-flies, they infest also some other trees and shrubs. The gall of the wild rose, Fig. 210, is very beautiful, being bright and variegated in color, and covered over with bristles. When cut open we find in

Fig. 210.—Gall of the Wild Rose.

Fig. 211.—Magnified Bristle of the Gall.

it little apartments occupied by the larvæ, as you see in the figure. In Fig. 211 you have one of the bristles magnified.

465. The insects of the Ichneumon family have long, slender bodies, long ovipositors, and long antennæ, which are in a continual trembling motion. The ovipositor of some species is exceedingly long, as in Fig. 212 (on the opposite page). The two bristles accompanying the ovipositor can be brought together by the animal so as to make a complete sheath for it. These insects deposit their eggs in the bodies of the larvæ of other insects, and the larvæ hatched from them live on these bodies just as the gall insects live on the galls in which they are hatched. Those which have long ovipositors pierce with them the bark of trees or decayed wood, in order to find larvæ in which they can deposit their eggs. Those which have shorter ovipositors deposit their eggs in the bodies of caterpillars which they find crawling about. We sometimes see a caterpillar with a considerable number of little barrel-shaped silken bodies standing out upon its skin. These are the cocoons of the Ichneumon larvæ, which, after living for some time in the fat of the caterpillar, just under its skin, have come out and have spun their cocoons, that they might go into the pupa state. The Ichneumon family is very numerous. Carpenter states that there are probably over three thousand species in Europe alone.

466. The Chrysididæ, or Ruby-tailed Flies, are a small group, adorned with such brilliant metallic tints that they have been said to be the Humming-birds of the insect world. The females deposit their eggs in the nest of wild Bees and other Hymenoptera, and thus the larvæ eat the food designed by these latter for their own offspring. Here is a striking analogy to the habit of the English Cuckoo, alluded to in § 268.

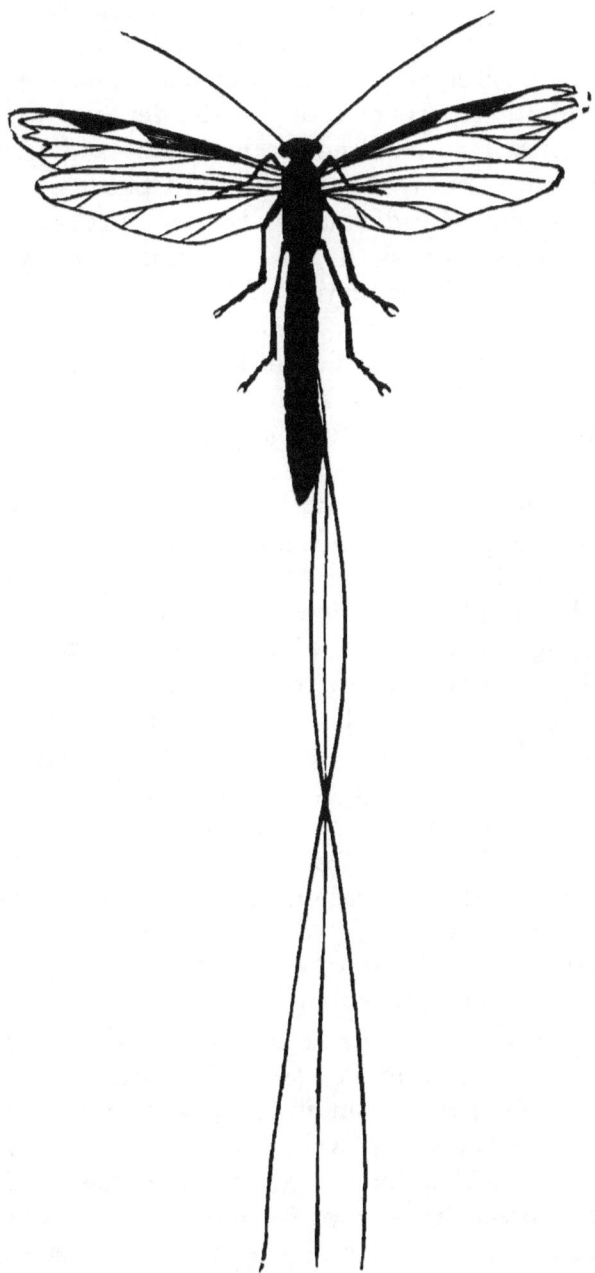

Fig. 212.—Long-tailed Ichneumon Fly.

467. The family of Sawflies is quite an extensive aberrant family. They are so called from a curious double saw in the ovipositor, with which they make holes in the branches and other parts of trees for the deposit of their eggs. Carpenter mentions one species in England, whose larvæ are very destructive to turnips, devastating a whole field in a few days by devouring the soft tissue of the leaves; and he states that the most effectual remedy has been found to be the introduction of ducks into the fields, as they very greedily devour the larvæ.

468. Of the Aculeate division of the Hymenoptera we make two subdivisions—the Predaceous, or those which live on prey, and the Melliferous, or honey-collecting stingers.

469. There is one group of the Predaceous division, including several families, which may be called, from their peculiar habits, diggers. They are known commonly as *Sand* and *Wood* Wasps. They are solitary—that is, do not live in communities. They therefore are all males and females, and have no neuters or workers. The females commonly dig out cells in the ground, or in posts and timbers. In these they deposit with their eggs insects which they have killed, so that the larvæ, when hatched, may have something to live upon. Sometimes the insects thus deposited are only stung sufficiently to render them powerless. Decomposition is thus prevented, and the larvæ, when they come forth from the eggs, kill the insects and devour them. The perfect insects are active in their habits, flying about and running over sand-banks with their wings in constant motion. They are fond of the nectar of flowers, a very different food from that which they devour in the greedy larva state. Those which are sand-burrowers have strong brushes on their legs with which they excavate their nests, while the wood-burrowers have powerful mandibles with tooth-like projections, which convert the wood into sawdust in making the burrow.

470. The Mud-wasp, Fig. 213, is one of the sand-burrowers. The following is the account given of it by Jaeger. "This insect is more than an inch long, and of a dark blue-purple color. It makes its abode in the loose, sandy ground, and when digging its hole resembles a dog digging after mice, throwing the earth under it toward its hind body with its fore feet. If the pile of sand becomes too high or troublesome, it places itself upon it, and throws the earth behind it with great force until it is leveled. As soon as its subterranean abode is prepared, it seizes a large Spider, or a caterpillar, or some other insect, stings it in the neck, and then carries it into its hole. It is curious to see one of these Wasps take hold of a Cockroach, seizing it by one of its long antennæ, and continually walking backward, compelling the Cockroach to follow, notwithstanding its great reluctance and constant opposition, until both have arrived at the hole, where the Wasp kills it by a sting in the neck, then tears into pieces, and carries it into her subterranean dwelling as food for her offspring."

Fig. 213.—The Mud-wasp.

471. The family of Vespidæ, or true Wasps, is distinguished from the other Hymenoptera by the folding of the wings when at rest throughout their entire length. They are generally not solitary, but social, the communities, however, being small. The neuters are not, like the neuters of the Ant tribe, destitute of wings. Those Wasps which are solitary have no neuters, and their habits are like the diggers just noticed. There are many

species of the Social Wasps, the best known of which, as the common Wasp, build their nests of a stout brown paper, which they manufacture from bits of wood and bark. Like the paper-maker among men, they reduce their material to a pulp, and then spread it out thinly, which, drying speedily, becomes firm paper. In Fig. 214 you see the arrangement of the nest of the Social

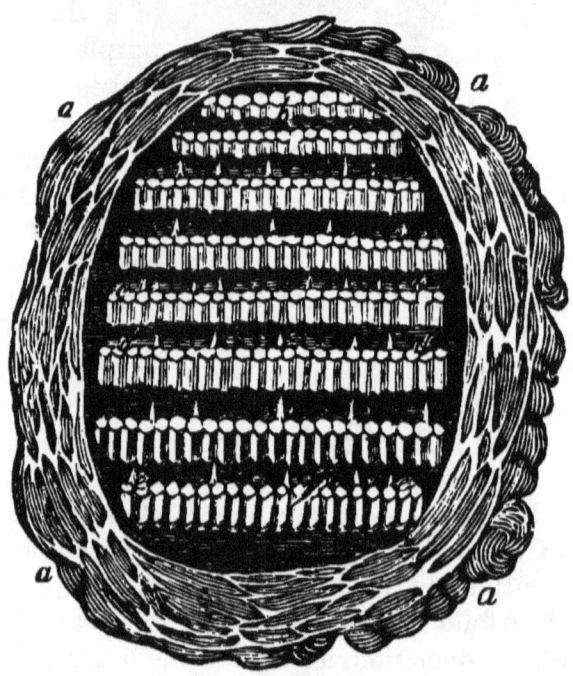

Fig. 214.

Wasps. Each floor of cells hangs from the floor above it by rods. At *a a* is the outer wall, made of many layers of brown paper; at *b* and *c* are five terraces of cells for the neuter Wasps; and at *d* and *e* are three rows of larger cells for the males and females. In Fig. 215 is a representation of a portion of one of these terraces, with its rod.

Fig. 215.

472. The family Formicidæ, or Ants, are placed in a different order from the

White Ants, § 453, on account of the difference in the wings, those of the latter having the characteristic network of the Neuroptera. They are distinguished from all the other families of the Hymenoptera by their residing under ground in large societies, some of them raising the earth up in mounds in constructing their habitations. The males and females, which alone are winged, constitute but a small portion of each community, most of it consisting of wingless neuters or laborers. The different parts of the nest are very curiously and regularly arranged. The males and females leave the nest as soon as they have wings. The males die, and of the females some return and deposit their eggs in their original nest, while others go to a distance and found other colonies. When they begin to lay their eggs, as their destiny is now to stay in one place, they have no farther need of wings, and therefore strip off themselves the useless encumbrances, or allow them to be stripped off by the neuters. These last not only construct the nest, but take care of the eggs, and also of the grubs that are hatched from them, feeding them, and carrying them on clear warm days to the outer surface of the nest, and taking them back again when night approaches, or before that if there be a threatening of bad weather. Ants are very fond of saccharine matter, and accordingly are apt to find out where it is. They are also fond of some fruits. I have been amused to see how any pear in my garden, that chances, in falling, to have a breach made in the skin, is at once beset with Ants, who quite rapidly eat out the inside.

473. In most cases a community of Ants consists only of three kinds of individuals—males, females, and neuters. But in some of the species some of the neuters are larger than the rest, and differently shaped, and appear to be the soldiers of the community, whose duties are the same with those of the soldiers among the Termites (§ 455). There are wars, sometimes, between different

communities among the Ants as well as among men, and some interesting descriptions have been given of their battles. But the most remarkable fact in the history of Ants is the propensity of certain species to kidnap the workers of other species and train them as their slaves. The kidnappers are always red or pale colored, while those which are made slaves are black. The slaves are not captured after they have become Ants, but when they are in the pupa state. The ant-heap is attacked by the marauders when the cells are filled with pupæ, and at no other time; a sanguinary battle is the consequence; and the Red Ants, being uniformly victorious, carry off the pupæ to their own nest. Here the red workers take the same care of these pupæ as they do of those belonging to their own community, and when the black workers come out from the pupa-cocoons, they very readily serve their captors.

474. The insects of the Melliferous or honey-collecting division of the Aculeata are distinguished by a peculiar conformation of the hind feet. The first joint has the shape of a square plate, on the inside of which are hollows surrounded by brush-like tufts. In these baskets, as they may be called, the pollen of flowers is collected and carried to the nest. The insects of this division are all called Bees. Like the Wasps, they have two groups, the Solitary and the Social. Of the Solitary Bees, some form burrows in the ground; others build several cells together, covering them with sand or small gravel united by their viscid saliva, and hence are called Mason Bees; others excavate cells in dead wood, and are called Carpenter Bees; and others still, the Upholsterer Bees, construct their nests from leaves, which they cut into the requisite shapes with great dexterity.

475. Of the Social Bees there are two principal groups —the Bombi or Humble Bees, and the Hive Bees. The Humble Bees, of which there are many species, build their nests either under ground, or on the surface under

stones and other things. Their communities consist of from fifty to three hundred Bees. They contain three kinds of individuals — males, females, and neuters, of which the females alone live through the winter. There is a special provision for their preservation. They have a chamber distinct from the rest of the nest, and this is lined with grass and moss, so that they may sleep there through the winter secure against the cold.

476. In the Hive Bees, we see the instincts in construction and social organization exhibited in the most remarkable manner. A community of Hive Bees contains but a single perfect female, termed the Queen; several hundred males, called Drones; and about twenty thousand neuters, or Workers. In Fig. 216 you see the three kinds of Bees, the upper one being the Queen, a drone on the right, and a worker on the left. At the

Fig. 216.—Hive Bees.

end of every summer, the Drones, which have no stings to defend themselves, are all stung to death by the Workers. The cells where the eggs are deposited are in the central part of each comb, this being, of course, the warmest part of it. The Drones being larger than the Workers, those cells in which the eggs are to produce Drones are made larger than the rest. Those that contain eggs from which Queens are to come are made much larger than any others, and of a different shape, as seen in Fig. 217. If left to themselves, the Bees select the hollow of an old tree as the place to build their comb, but they readily go into the hives provided for them by man. The comb is made of wax, which is secreted from the body of the Bee, and appears in scales between the segments. This, with their mandibles, they mould, and apply it in the construction of the comb. The first swarm from a hive is led by the old Queen. Then there are successive swarms led by the young Queens. When the hive is sufficiently relieved of its surplus population, the Queens that remain fight till all are killed but one, and she takes possession of the throne. When the Bees lose their Queen, they, by a curious process, fill the vacant throne. They select some one of the larvæ which are to be Workers, and, enlarging its cell by the removal of the walls of the neighboring cells, thus make for it a suitably large royal apartment. They then feed it with the *royal jelly*, which is the exclusive food of the Queen-larvæ, and, in due time, this larva, originally destined to be a worker, comes forth a queen.

Fig. 217.—Royal Cells.

Questions.—What is said of the wings of the Hymenoptera? What peculiarity is there in the females? What is said of the instinct of the Hymenoptera? What is said of their metamorphosis? What of their larvæ? What is said of the size of the Hymenoptera? Of their number? Of their habits? Of their usefulness to man? What are the two groups of this order? What is said of the Gall-flies? What of the Oak-galls? What of the gall of the wild rose? What are the peculiarities of the Ichneumon family? What is stated of those which have short ovipositors? What is said of the Chrysididæ? What of the Saw-flies? What are the families of the aculeate group of the Hymenoptera? What is said of the Sand and Wood Wasps? What of the Mud-wasp? What of the Vespidæ? What is the arrangement of the nest of the social Wasp? Describe the Ant communities. What is said of the neuters in some species?

What of the kidnapping Ants? What are the peculiarities of the Melliferous division of the Aculeata? What is said of the solitary Bees? What of the social Bees? What of the hive Bees? What of their swarming?

CHAPTER XXVIII.

SCALE-WINGED INSECTS.

477. The insects of the order Lepidoptera, or Scale-winged Insects, are characterized by the downy covering of the wings, which is made up of a multitude of feather-scales. The number of these scales on the wings of the Silkworm Moth has been estimated at 400,000. The silvery dust that you have on your fingers when you touch a common Miller is a multitude of these scales. Each particle of that dust under the microscope appears a scale, with regular lines extending from its stem to its edge at the other end. When this scaly covering is rubbed off from the wing of one of these insects, the bare membrane which is left is seen to correspond with that of the wings of other insects. In some cases the scales are arranged with perfect regularity, § 398. The shapes of them vary much in the different species, and there is often quite a variety in the same species in different portions of the wing, the long ones making the fringe at the edge. That you may have a correct idea of their general shape, I give, in Fig. 218.

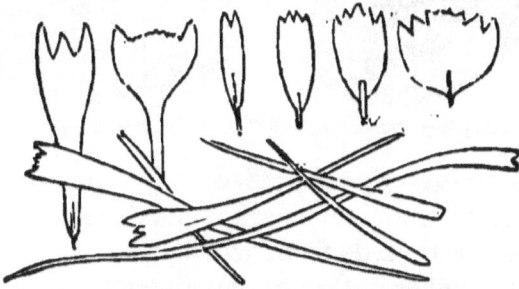

Fig. 218.—Feather Scales of the Goat-moth.

the feather-scales of the Goat-moth. The delicate lines on them are not represented. It is these scales variously colored that give such beauty to many of the insects of this order. Some of the Butterflies are especially brilliant.

478. The insects of the orders already noticed are mandibulate, § 392. This order, and the others which remain to be noticed, are haustellate, § 393. The Lepidoptera stand at the head of the haustellate group, as the Coleoptera, or Beetles, stand at the head of the Mandibulata. The haustellum, or sucker, by which the insect drinks up the nectar of the flowers, is composed of two long filaments, so shaped that, by joining them together, they make a tube. You can see how accurately they must be made in order to do this.

479. The larvæ of the Lepidoptera are caterpillars. They have three pairs of legs on the first three segments of the body; then they have some appendages called *pro-legs*, which are thick, short, fleshy tubercles, with minute hooks around the edge of the under surface of them: there are usually five pairs of these, four of them in rear of the true legs, and another pair on the last segment of the body. In Fig. 219 are represented a leg

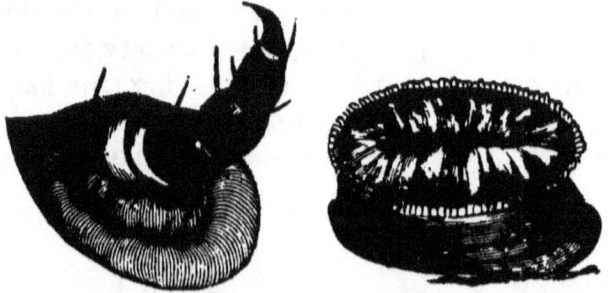

Fig. 219.—Leg and Pro-leg of a Caterpillar, greatly magnified.

and a pro-leg, greatly magnified. The curved claws on the six legs of the caterpillars enable them to climb up readily on the threads from which they so often hang, and the pro-legs are of great assistance to them in walking and in adhering to branches of trees, or to any other solid substance. Besides the little hooks on the pro-leg, the bottom of its foot is so arranged as to act as a sucker. The mode of walking or crawling is different in different caterpillars. Those which have pro-legs on nearly all the segments crawl on all the feet at once, moving the body straight along. Those which, on the other hand, have only a few pro-legs, manage in this way: making firm hold with their six clawed legs, they bring the pro-legs, which are at the other extremity of the body, close up in rear of the true legs, thus arching the intermediate segments upward; and now, holding on with their pro-legs, they thrust the anterior part of the body forward its full length. By a repetition of these movements they make a slow and measured progress. From this mode of walking such caterpillars are called loopers, or geometers, or measure-worms. Some caterpillars will stand for hours on the pro-legs in the rear part of the body, with the forward part of the body extending upward at right angles to this rear part.

480. The food of caterpillars is, with few exceptions, vegetable. Some feed exclusively on one kind of plant, as the Silkworm on the mulberry; some feed on a certain class of plants; and others on almost any kind that they happen to find. Their hours of eating differ, some eating only in the morning and evening, some all day, and others only at night. All eat a great deal—some more than twice their weight in twenty-four hours. If all animals should do this, the eatables in the world would soon be devoured. The perfect insects eat but little, for they do not grow any larger than they were when they first emerged from the pupa state. The larvæ, on the other hand, as they eat much, grow much also.

481. Caterpillars are of great service in furnishing a very large proportion of the food of birds. "It is ascertained," says Jaeger, "that a single robin or woodpecker, and many others of the warblers, carry every day about

fifty grubs or caterpillars to their nests as food for themselves and their young. Now, if there were only one million of these birds, of which each one devours 6000 caterpillars in the months of April, May, June, and July (by no means a large computation), the number of caterpillars and grubs thus destroyed will amount to 6,000,000,000 annually."

482. Caterpillars are all spinners, the thread coming from a fleshy point in the under lip. Besides the employment of this spinning machine in making the cocoon for the pupa state, many of them also use it as a means of escape from their enemies, letting themselves suddenly down by the thread they spin to a place of safety. If a bird espies one in a rolled-up leaf, he may not secure it, for, as he puts in his bill at one end, the caterpillar may escape at the other, dropping itself down quickly as far as it pleases.

483. Most of the caterpillars are solitary in their habits, but some live in societies. This is the case with the Tent-caterpillars. These spin large tents of silk in the branches of trees, which are water-proof, although they are so slight in their appearance. They increase very fast, and, if let alone, colonies from the original community will spread their web-like tents in all parts of the tree.

484. Of the caterpillars called Spanworms there are many species. The most conspicuous is what is commonly called the Canker-worm, so destructive to many fruit and shade trees from devouring their leaves. These caterpillars finish their work of devastation in June, when they are only four weeks old, and descend by their silken cords to the ground, which they enter to the depth of several inches. Here they pass into the pupa state. In the autumn they issue from the ground in the imago state. The female is wingless, and therefore must climb up the trunk of the tree to lay her eggs on the branches, which she does in clusters of a hundred or more. There

they remain till spring, when the caterpillars are hatched from them. Various expedients have been devised for destroying the females as they go up the trees to lay their eggs. The most effectual one is that adopted so extensively in New Haven to save its noble elms. It is a leaden trough placed around the trunk of the tree in which there is some kind of oil.

485. Professor Jaeger very playfully says of the habits of the perfect insects of this order that, "in comparison with the other orders of insects, they are well entitled to the rank of nobility, for among them we find no impudent beggars and spongers, as among the Flies; no parasites, as among the wingless insects; no working-class, as among the Hymenopterous insects—Bees, Wasps, and Ants; no musicians, as among the families of Crickets, Grasshoppers, Katydids, and Cicadas; but all of them are aristocratic idlers, who, clothed with silver, gold, and purple, and ornamented with ever-varying splendor, have naught to do but seek their own pleasure, and charm away their brief existence, fluttering from bough to bough, and satiating themselves with the sweet nectar of flowers."

486. We divide the Lepidoptera into two sections— the Butterflies and the Moths. The Butterflies may usually be distinguished by the vertical position of their wings when they are at rest, and by their having the antennæ slender, and club-shaped at the end. They are diurnal in their habits, and they are therefore brilliant, generally, in their colors. The under side of the wings is as beautiful as the upper. The pupæ of many of this group have golden spots, from which the term chrysalis was suggested, and also aurelia, which is a Latin word of the same meaning with the former, which is Greek. These terms ought strictly to be applied only to the pupæ of Butterflies, but they have come to be applied to pupæ of all kinds.

487. The Butterflies are divided into five families, ac-

cording to the shape of the wing. One of these families, styled by Linnæus Knights or Chevaliers, generally have a long, swallow-like tail at the extremity of the hind wings, as seen in the Troilus, Fig. 220. This Butterfly

Fig. 220.—The Troilus.

has black wings spotted with yellow. Its caterpillar is green, with a yellow stripe on each side, and a row of blue dots, while the under side of its body and its feet are reddish. In this country it is more frequently seen in the Southern than in the Northern States. The Butterfly called Berenice, Fig. 221, belongs to the family of Round-winged Butterflies. It was named after the wife of Antiochus, King of Syria, said to be the loveliest woman of her age. It is quite common with us. It has dark-red wings with black veins, and a black border with two rows of white dots. The caterpillar is of a light violet color, with brown, red, and yellow lines.

488. The second section of the Lepidoptera, the Moths,

Fig. 221.—The Berenice.

we divide into two groups, the Crepuscularia (*Crepuscula*, twilight), Twilight-fliers, or Hawk Moths, and the Nocturna, or True Moths. Linnæus called the Hawk Moths Sphinxes, from the peculiar attitude, resembling the sculptured Sphinx, so often assumed by the caterpillars of these Moths. Most of the species in this genus are Twilight-fliers, but not all; for some fly about in bright sunlight, sucking the nectar of flowers with their long trunks. These species are more brilliantly colored than the common species, which have a dull brownish-gray aspect, like the owls, whose habits are similar. The larvæ of the Hawk Moths, on going into the pupa state, either inclose themselves in cocoons, or bury themselves in the ground. The perfect insects make a loud humming sound in their flight. The Humming-bird Moth is one of the most beautiful of the diurnal species, and is remarkable for the loudness of the humming sound which it makes while feeding poised on its wings.

489. The Nocturna, or True Moths, are by far the most extensive group of the order. They are much like most of the Sphinxes, but their antennæ are very different, being broad at the base, and tapering to a point at the end. The Cecropia, Fig. 186, is one of the most splendid of these Moths. The Silkworm Moth belongs to this group; so do all that variety of Moths, or Millers, that fly about

our lights in a summer's evening. I have said so much of these insects in Chapter XXIII., that of the many families of them I will notice here but two. The Clothes Moths deposit their eggs in woolen stuffs, furs, feathers, etc. Their larvæ live on these articles. They also construct for themselves a tubular case from the same materials. In these they live, as the larvæ of the Caddice-flies, § 459, do in their cases. With the growth of their bodies they enlarge these tubes by weaving an addition on to the end, and also by slitting it open and inserting a piece longitudinally. Sometimes these cases are of divers colors, from the use of differently-colored materials. When they are about to go into the pupa state, these insects close up the two ends of the case.

490. There is a small Moth, called the Rusty Vapor Moth, Fig. 222,

Fig. 222.—The Rusty Vapor Moth.

of a light-brown color. Though it is rather homely, it comes from a caterpillar which is very beautiful, represented in Fig. 223. Its body is covered with long, fine yellow hairs, and has at each end two elegant brush-like tufts. Its head is as red as sealing-wax, and there are prominences on its

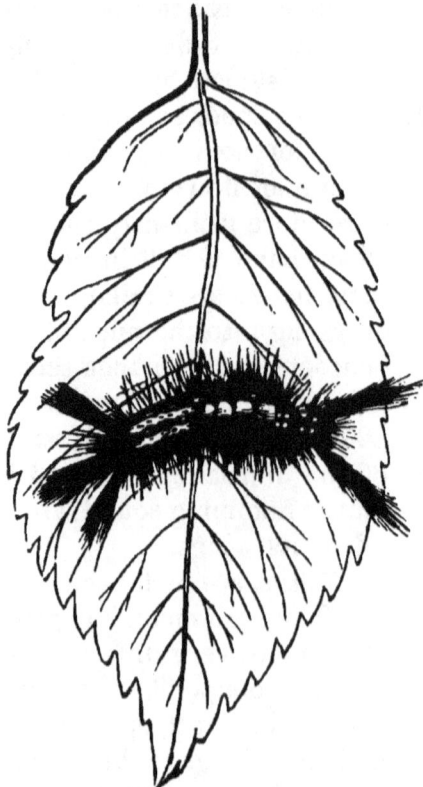

Fig. 223.—Caterpillar of Rusty Vapor Moth.

back of the same color. The motions of these caterpillars are very slow, and they eat but little. Commonly the perfect insect has more beauty than the larva from which it comes, but here we have an example of a contrary character.

Questions.—What characterizes the wings of the Lepidoptera? What is said of the shapes and arrangements of the scales? Which are the Mandibulate orders of insects, and which are the Haustellate? What is the construction of the haustellum of the Lepidoptera? What are the two kinds of legs of their larvæ, and how are they used? Describe the two modes of walking. What is the food of caterpillars? What is said of the quantity which they eat? How are caterpillars of great service to us? What is said of their spinning? What is said of the Tent-caterpillars? What of the Canker-worm? What is said of the habits of the Lepidoptera? How are the Butterflies distinguished from the Moths? What is said of the pupæ of the Butterflies? How many families are there of Butterflies? What is said of the Troilus? What of the Berenice? What are the two groups of Moths? What is said of the Crepuscular Moths? What of the Humming-bird Moth? What of the Nocturnal Moths? What of the Clothes Moth? What of the Rusty Vapor Moth and its larva?

CHAPTER XXIX.

HALF-WINGED AND TWO-WINGED INSECTS.

491. THE insects of the order Hemiptera present many curious varieties. They agree, however, generally in the arrangement of the mouth, it being adapted to suction by a beak which is singularly constructed. It is a horny sheath, containing in a channel or groove four stiff bristles as sharp as needles. This instrument, which is thus fitted for both piercing and sucking, when not in use is bent under the body, and lies against the chest. This order is termed by some Rhynchota, from a Greek word, meaning beak. The food of these insects consists of the juices of plants in most cases, but in some of those of animals. They are called Hemiptera, half-winged, on

account of the peculiar construction of their wing-cases, the fore part of which is thick and opaque, while the hinder half is thin and transparent. There are some which have the wing-covers transparent throughout, and some, also, that have no wings—as Bedbugs; but, as both have the peculiar beak of this order, they are ranked here.

492. The insects of this order do not appear first as caterpillars, like the Butterflies; or as grubs, like the Beetles; or as maggots, like the Bees and Flies. They come forth from their eggs in an almost perfect condition, except that they are then wingless. The Cicadas, however, are an exception. They live in the larva state in the ground even for some years. I will notice a few of the prominent families of the order.

493. Of the family of Cicadæ, famous for their chirping sounds, the Red-eyed Cicada, or Seventeen-years Locust, Fig. 224, is the one with which we are familiar.

Fig. 224.—Red-eyed Cicada.

The females deposit their eggs on the trees; the larvæ hatched from them descend and enter the ground, where they feed on roots. The change from larva to imago is effected in this way. When this is about to take place, the grub comes up out of the ground, and, with its strong feet, fastens itself to a fence or the trunk of a tree. The back now gapes open, as seen in Fig. 225 (p. 291), and a winged insect comes forth, leaving the horny shell of its grub state clinging to the spot where the change

HALF-WINGED INSECTS.   291

Fig. 225.—Grub of Cicada.

takes place. Sometimes the animal is not able to effect its exit, and dies in the struggle. These shells may often be found clinging to trees and fences in considerable numbers. It is supposed that the Seventeen-years Locusts really remain in their grub state under ground seventeen years, but Jaeger holds to the contrary.

494. The Frog-hoppers are so called from their great power of leaping. Those of this family most familiar to us are the Tree-hoppers, of which a specimen is given in

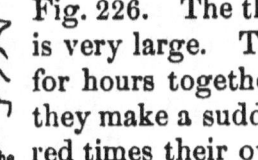

Fig. 226.—The Tree-hopper.

Fig. 226. The thorax or chest of these insects is very large. They are commonly motionless for hours together; but if they are disturbed, they make a sudden leap of two or three hundred times their own length, and, spreading out their wings, fly off to some other spot. The insects of this family are sometimes called Froth-hoppers, from a frothy fluid which exudes from them. In some species, in tropical countries, this exudation is very abundant.

495. The Aphidæ, or Plant Lice, have small, round, full bodies, presenting different colors on different plants. Some have wings and some have not. They live in great numbers on the stalks and leaves of plants, sucking the sap with their beaks. The postures which they sometimes assume is very amusing. I saw the past summer in my garden some stalks of the wild Aster lined with them from top to bottom, and every one had its head downward. The hind legs did not touch the stalk at all, but were raised up, and the insects rested on the fore legs and the beak. Thus standing out, and being of a reddish color, they gave the appearance of ornamental appendages, until the eye was brought near enough to see what they were.

496. On the back of these insects there project behind two tubes, from which issues a sweet fluid. Ants are very fond of this, and take it from the tubes as it exudes, or from the surface of the plants, where it is known as *honey-dew*. The Aphides are, therefore, appropriately called the milch-cows of the Ants. Some species of Ants even gather them into flocks, and keep them in a sort of pasture, as we do cows.

497. The Scale-insects, though very small, are, like the Aphides, greatly injurious to plants. Like them, they are abundantly prolific, and when they once get possession of a plant or young tree, it is almost certain to die, the minute size of the larvæ of the insect rendering it almost impossible to find and exterminate them. The name Shield-louse, so often given to these insects, is derived from the appearance of the female, which, with its shield-shape, clings tightly to the plant, looking more like a wart than an animal. It lives on the sap, which it sucks with its beak or snout. It deposits eggs on the bark, covering them with a sort of cottony secretion. It then dies, and its dried body forms another covering for the eggs. The cochineal, so valuable to commerce, is a scale-insect. It is found chiefly in Mexico and Central America. It is estimated that the export of cochineal from these countries is to the amount annually of two and a half millions of dollars. This rich dyeing material was used for a long time without its being known what it was; and a French naturalist, in 1792, was universally ridiculed for asserting that cochineal was an insect. It is gathered from cactus plants, which are largely cultivated in plantations for the purpose of raising this insect for the market. The lac of the East Indies, so extensively employed in the composition of varnishes, sealing-wax, etc., is the product of another species of these insects.

498. There are various bugs belonging to this order, in some of which the wings are entirely absent. They

are divided into two sections—Land-bugs and Water-bugs. To the former section belong the Bedbugs before referred to. Of the Water-bugs there are only two families—the Boat-flies and the Water Scorpions. The former are good swimmers, always swimming on the back. They can fly well, but rarely do it.

499. The Diptera, or two-winged insects, constitute one of the most extensive orders, both in the number of species and in the number of individuals. None of them are large, and some are exceedingly small. For the most part they are dull in color. On the head are two very large compound eyes, and two short antennæ near together. In some there is a soft proboscis, as the common House-fly; in others, a hard, pointed, sucking tube, as in the Musquito; and in others still, simply a mouth. They have three pairs of feet, and two thin wings, which, in most cases, give out a humming sound in flying. Their larvæ are generally maggots, white, and having no feet, but instead thereof fleshy tubercles or warts, on which they crawl. Most of the larvæ live in dirt, or dung-hills, or spoiled meat, or cheese, etc. The metamorphosis is complete, but in some cases very peculiar.

500. The species of Flies are very numerous. There are about seventeen hundred known in Europe. The larvæ of Flies, the maggots, generally live in some kind of filth; but the Flies themselves live, for the most part, on dainty food. The wing of a common Fly, examined under the microscope, is a beautiful object. Although to the naked eye it has a very plain appearance, it is covered with little pointed projections of curious shape regularly arranged.

501. The larvæ found in cheese come from eggs deposited by a small Fly. From their great power in leaping they are called Cheese-hoppers. The manner in which the leap is performed is very singular, and is thus described by Carpenter: "When preparing to leap, it first raises itself upon its tail, in which position it is enabled

to balance itself by means of some prominent tubercles on the last segment of the body. It then bends itself into a circle, and having brought the head toward the tail, it stretches out the two hooks of the mouth, fixing them into two cavities at the other extremity of the body. It then contracts the body from a circular to an oblong figure—the contraction extending in a manner to every part of the body. It now suddenly lets go its hold, and straightens the body with such violence that the noise produced by its hooks is very perceptible. The height of the leap is often from twenty to thirty times the length of the body, exhibiting an energy of motion which is particularly remarkable in the soft larva of an insect. A Viper, if endowed with similar powers, would throw itself nearly a hundred feet from the ground."

Fig. 227.—Wriggler.

TWO-WINGED INSECTS. 295

502. The Musquito family are remarkable in many respects, but chiefly for the peculiar mode of their metamorphosis. The common Musquito, when first hatched, is an inhabitant of the water, and is, from its antic and rapid motions, called a Wriggler. In Fig. 227 you see the animal of its natural size, and also as it looks when magnified. Though it lives in the water, it is not like a fish, for it has no gills. It is more like a whale, for it is obliged to come occasionally to the surface to breathe. Its breathing apparatus is near its tail. The air is taken in through a tube made of hairs, represented at A. After the insect arrives at its proper size it comes to the surface with its back upward, which gapes open, as in the case of the Cicada (§ 493), and the winged insect emerges, as seen in an enlarged representation in Fig. 228. It

Fig. 228.

rests upon its cast-off skin as a boat, while it unfolds and expands its wings, and then flies off. Great care is required in this operation, as there is danger that the insect will be plunged into the water before it expands its wings.

503. The eggs of the Musquito are deposited on the surface of stagnant water to the number of about three hundred, fastened together as you see in Fig. 229. They thus make a sort of raft which swims on the surface.

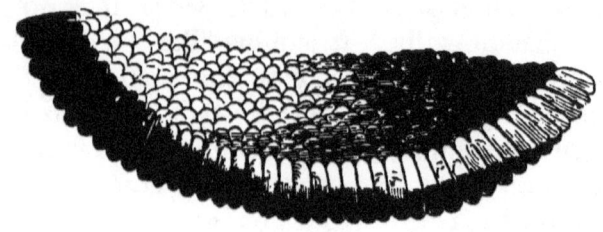

Fig. 229.

The large ends of the eggs are downward, and it is out of these that the larvæ come, diving down into the water. There is a lid at the blunt end of the egg which is opened to let the larva out. Some species do not have this mode of arranging their eggs.

504. The proboscis which is visible to us, and which the insect so deliberately adjusts upon the skin when it alights, is not the stinging apparatus, but the sheath or scabbard of it. It incloses some bristles with lancet-shaped points. When the skin is pierced by these, the blood is sucked up through the sheath. It is supposed that the irritation attending the bite is occasioned by the saliva of the insect introduced into the wound to dilute the blood that it may more readily be sucked up. In Fig. 230 you have at A the sheath closed, both of the natural size and magnified. In the lower figure you have the whole instrument opened — at B the sheath, at C three lancets, and at D protectors. At F you see these parts of their natural size. This is the arrangement of the proboscis of the common American Musquito. It is different, however, in the different species of this insect.

505. The different species of Musquitoes, of which there are many, are quite widely diffused in the earth.

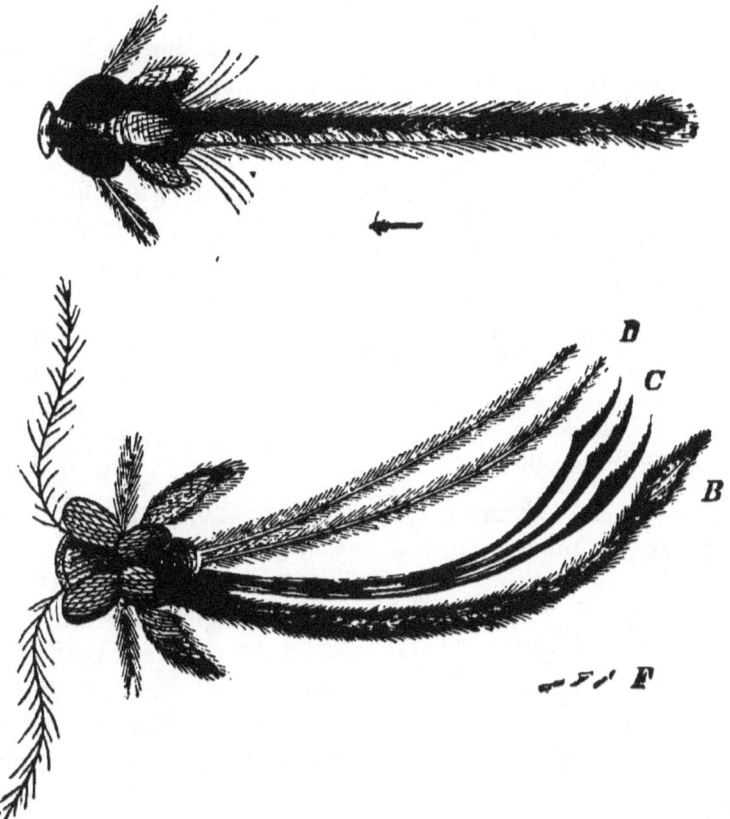

Fig. 230.

They are generally most troublesome in warm climates, and in the tropics they are present throughout the year. But there are some cold countries in which, during their brief but hot summers, they are not only extremely annoying, but occasionally very destructive. This is the case with parts of Russia, both in Europe and in Asia. Even such animals as horses, oxen, sheep, goats, and hogs, are so severely stung by them as to die, some meeting their death by drowning, having run into water to escape the swarms of their small but formidable enemies. At some periods it seems in that country to be the grand business of life to devise and put in execution expedients for guarding man and beast against these insects.

506. The insects of the order Aphaniptera, the Fleas and their allies, have only the most indistinct rudiments of wings; but the metamorphosis is complete. The larvæ inclose themselves in small silk cocoons to pass into

Fig. 231.—Flea.

the imago state. The common Flea, a magnified representation of which is given in Fig. 231, has a curious apparatus for sucking blood, which is very beautiful as examined with a microscope. This insect, like other great leapers, as Grasshoppers, Frog-hoppers, etc., has very large hind legs.

Fig. 232.—Louse.

507. In the order Aptera, or wingless insects, are found the different kinds of Lice which infest different animals. In Figure 232 is represented the common Louse; at a of the ordinary size, at b magnified. At c is one of its legs magnified; at d are its eggs, also magnified.

Questions.—What is said of the haustellate apparatus of the Hemiptera? What gives them their name? What is said of their metamorphosis? How are the Cicadas an exception to this? What is said of the Red-eyed Cicada? What of the Frog-hoppers? Describe the Aphidæ and their habits. What is said of their honey-dew? What is said of the scale-insects? What of Cochineal? What is said of some aberrant bugs of this order? What is said of the extent of the order Diptera? What of the size of these insects? What are their peculiarities? What is said of their larvæ? What is said of Flies? What are Cheese-hoppers? Describe their mode of leaping. Describe the larvæ of the common Musquito. Describe

Its metamorphosis. What is said of its eggs? Describe the arrangement of the proboscis. What is said of the Musquitoes in various regions of the earth? What is said of the order Aphaniptera? What of the order Aptera?

CHAPTER XXX.

THE ARACHNIDA.

508. THE second class of the Articulata is that of the Myriapoda, the Centipedes, § 388. This I will not dwell upon, but will pass directly to the third class, the Arachnida. This class was for a long time included among the Insects, and Spiders are very generally spoken of now, in common conversation, as belonging to that class; but the Arachnida differ from Insects in several important particulars. The head of Insects is distinct from the chest, but in the Arachnida the head and chest are united in one; and this is called the cephalo-thorax. Insects in their perfect state have but six legs, but the Arachnida have eight. The Arachnida have not the compound eyes of Insects. Again, the antennæ of Insects are wanting in the Arachnida.

509. The Arachnida are carnivorous; but generally, instead of eating their prey, they suck the juices from their bodies. Many of them have a poison apparatus, by which they can destroy more readily those victims whose strength would otherwise be too much for them. They have mandibles and pincers very much like those of insects. In those which are parasitical—that is, those which dwell on other animals—the mouth has the form of a trunk or proboscis armed with a kind of lancet. The Scorpions have a curved and pointed instrument at the end of the tail, as seen in Fig. 233 (p. 300). They have large claws, like those of the Lobster, with which they seize their victims, and then pierce them with this curved sting, which is armed with poison from a gland.

Fig. 233.—The Scorpion.

510. The class of Arachnida is divided into two groups. In the first group the respiratory organs are different from those of Insects. Instead of passages every where for air, there are some sac or bag like cavities in the abdomen, and in these are thin membranous plates arranged like the leaves of a book. The air goes in among these, and acts on the blood in the vessels spread out on them. This group includes the Spiders and Scorpions. In the second group the respiratory apparatus is like that of Insects. This includes Mites of various kinds, Father-long-legs, the minute red Spiders of green-houses, etc.

511. Most of the true Spiders are great spinners. They do not spin for themselves a cocoon as the caterpillars do, for they undergo no metamorphosis. They spin chiefly for two purposes—to construct a dwelling for themselves, and to construct traps to catch their prey. Some also, like some of the caterpillars, spin as they drop to escape their enemies, and thus save themselves from a fall. Some throw out a long thread into the air from their spinning machine, and let it, when it is of sufficient length, bear them aloft like a balloon. And some spin a cocoon in which they deposit their eggs. I found one of these cocoons the past summer fastened to the bark of a tree. I opened it, and it was all a moving mass within. On looking at it with a pocket microscope, I found that it was full of little Spiders, which probably had just been hatched from the eggs, but were not yet ready to come out. The manner in which the cocoon is formed and filled with eggs is curious. The Spider first spins the lower half of it, and into this silken cup it drops the eggs. It not only fills it, but piles up eggs on top

with great care, so that there are as many above as in the cup. It then finishes spinning the cocoon.

512. The Caterpillars spin from the head, but the Spider spins from the other extremity of the body. Its spinning apparatus is of peculiar construction. Inside is a reservoir of gummy matter from which the silk is made. The threads of a Spider's web are drawn out from it, and dry as fast as they are drawn. But the thread, which appears to the eye as single, is found by the microscope to be composed of many thousands of threads united together. In Fig. 234 you see, as the Spider hangs by his thread, that it comes out from a circular spot. In this are four and sometimes six knobs, which can be seen by the naked eye. Each of these is full of holes through which the threads come, and these holes are so minute that Reaumur calculated that a thousand occupied a space no larger than the point of a pin. In Fig. 235 (p. 302) is represented such a view of these knobs as you would get by a powerful microscope. A portion only of the minute threads are represented. It was the calculation of Leuwenhoeck that it would take four millions of them to make a thread as large as a hair.

Fig. 234.

513. These threads are united together about one tenth of an inch in distance from the spinnerets. By this separate exposure to the air of each threadlet, they all become dry before their union. Another advantage of this

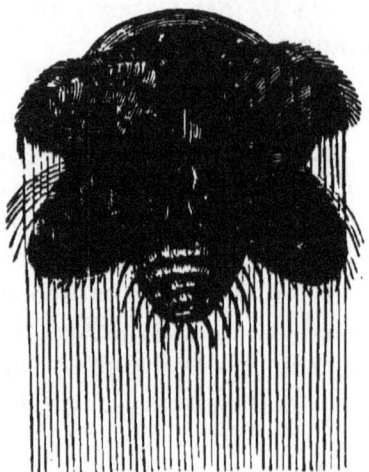

Fig. 235.—Spider's Threads coming from the Spinnerets.

arrangement is the securing of greater strength to the thread, for it is well known in rope-making that, in cords of equal thickness, those which are composed of many smaller cords are stronger than those which are spun at once. Another advantage still is, that these minute threadlets can be better attached to an object than a single thread. When the Spider makes an attachment of his thread, he presses the spinnerets against the spot selected, and thus fastens the ends of the threadlets projecting from the holes over quite a space. This is seen in Fig. 236, which represents an attachment of this kind, as seen with the microscope.

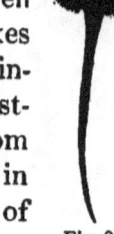

Fig. 236.

514. The foot of a Spider, a magnified view of which is given in Fig. 237, has three claws, one of which acts as a sort of thumb, and the others are toothed as a comb. It is supposed that these combs are used in preventing tangling of the threads in the web, and also in removing any particles that may become attached to it. When a Spider has let itself down from any place by its thread, if it goes up again upon it, it gathers up the thread into a ball with its claws and throws it away. So, also, if any part of its web is rendered useless by any thing which becomes attached to it, it is separated from

Fig. 237.—Triple-clawed Foot of a Spider, magnified.

the rest, collected into a packet, and cast off. Mr. Rennie, the author of a very interesting book on insect architecture, describes a process of this kind which he observed on board of a steam-boat. It was a geometric Spider, that is, one that forms its web of regular circular lines. The web or net was covered with flakes of soot. "Some of the lines," he says, "she dexterously stripped of the flakes of soot adhering to them; but in the greater number, finding that she could not get them sufficiently clean, she broke them quite off, bundled them up, and tossed them over. We counted five of these packets of rubbish which she thus threw away, though there must have been many more, as it was some time before we discovered the manœuvre, the packets being so small as not to be readily perceived, except when placed between the eye and the light. When she had cleared off all the sooted lines, she began to replace them in the usual way."

515. Many observations have been made, and experiments tried, to determine how Spiders transport themselves from tree to tree, across brooks, or even sometimes through the air, without any visible starting-point. The subject is not entirely cleared up, but it is well ascertained that they spin out the thread, letting the wind take it, trying it occasionally with the feet to decide whether the farther end has attached itself to any object. So soon as the Spider finds by pulling on it that it is fastened, it runs along upon it, strengthening its cable by spinning another as it goes. Spiders have not, as some have supposed, the power of projecting their lines in opposition to the moving air, but they uniformly put their bodies in such position that the line may go with the air, that is with the head toward the direction from which the breeze comes. They watch the wind as much as the sailor does. The little gossamer Spiders let their lines, like balloons, carry them off into the air, breaking loose from the objects on which they stand when they

feel themselves acted upon by a force sufficient for that purpose. They may thus be seen mounting aloft from the tops of twigs and blades of grass, from fences, etc.

516. The architecture of Spiders has considerable variety. That of the house Spider and that of the common geometric Spider are familiar to every one. That of the labyrinthic Spider is very curious. Its nest may be seen spread out a broad sheet on hedges, furze, low bushes, and sometimes on the ground. "The middle of this sheet," says Rennie, "which is of a close texture, is swung, like a sailor's hammock, by silken ropes extended all around to the higher branches; but the whole curves upward and backward, sloping downward to a long funnel-shaped gallery which is nearly horizontal at the entrance, but soon winds obliquely till it becomes quite perpendicular. This curved gallery is about a quarter of an inch in diameter, is much more closely woven than the sheet part of the web, and sometimes descends into a hole in the ground, though oftener into a group of crowded twigs or a tuft of grass. Here the Spider dwells secure, frequently resting with her legs extended from the entrance of the gallery, ready to spring out upon whatever insect may fall into her sheet-net."

517. There are some species of spiders that build their nests of clay, which they knead into due shape, and hence are called Mason Spiders. There is one of these found in the West Indies. This Spider digs a hole obliquely in the earth about three inches deep and one inch in diameter, the walls of it being made of clay. This cavity it lines with a thick web, which, when taken out, resembles a leathern purse. This tapestried chamber has a very singular door. It is made of about a dozen layers of this same lining, closely united together, and has a hinge of the same material. In Fig. 238 (p. 305) is represented the nest of another Mason Spider found in France, A being the nest shut, and B the nest open;

THE ARACHNIDA. 305

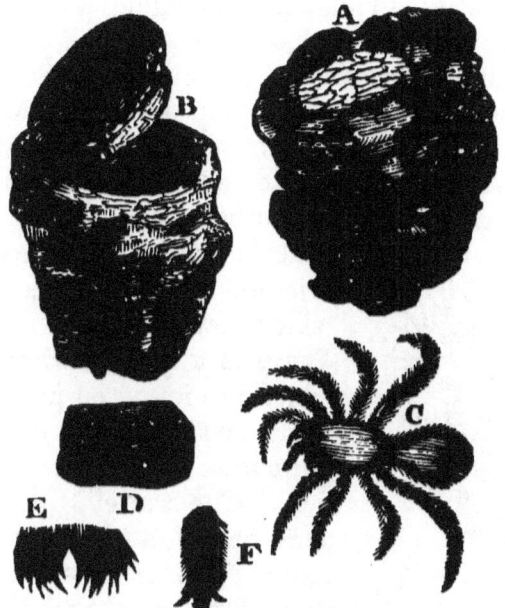

Fig. 238.—Nest of a Mason Spider.

C the Spider, D the eyes magnified, and E and F parts of the foot and claw magnified.

518. There is a Spider common in the woods that weaves together a great many leaves for a dwelling, and in front of this spreads its snares to catch its prey. When winter approaches it leaves its eggs in this nest to be hatched the following spring, and itself retires to some hollow tree to die.

519. An English clergyman, Mr. Shepherd, has often seen in the fen ditches of Norfolk a very large Spider that makes a raft by fastening weeds together with silken threads, and sails forth on this in search of insects that may chance to get into the water. But the most interesting water-spider is one that makes for itself a silken diving-bell, which looks in the water like a little silver globe. This is sometimes partly above the surface of the water, but at others it is fastened by silken ropes to objects below. The Spider contrives in some way to carry

air down to its diving-bell, coming up every now and then to the surface for this purpose.

520. I have already said enough of the Scorpions (§ 509), and on the second group of the Arachnida I will spend but a few words. Among the Mites is the animal which occasions the disease called the itch, an enlarged representation of which you have in Fig. 239. It has an

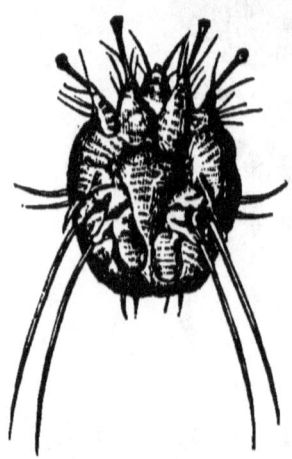

Fig. 239.—Sarcoptes Scabiei, or Acarus of the Itch.

oval body, a mouth armed with bristles, and eight feet, four of which have suckers at the end. There is a great variety of mites which are found on plants and animals, and some live in the water, swimming about with great freedom. The scarlet Mite of our gardens has a most brilliant scarlet color. The Harvest-men, so appropriately called Father-long-legs, as they have, perhaps, longer legs than any other animal of any kind, are mostly very agile. The Book Scorpions, so called, are little Arachnida which inhabit herbariums, old books, etc. They are good runners, often going sidewise like crabs, and they hunt the minute insects which are found in such situations.

Questions.—How do the Arachnida differ from insects? What is said of their food? What of their means of killing their prey? What is said of those which are parasitical? What is said of the Scorpions? What are the two groups of the Arachnida? What are the two chief purposes for which Spiders spin? What other purposes are sometimes accomplished by it? What is said of the cocoons which some Spiders spin? Describe the spinning apparatus of Spiders. What is said of the compound character of the Spider's thread? Why is it not spun whole? What is said of the mode of its attachment? Describe the foot of a Spider. What is the use of the combs in it? Describe its mode of repairing its web. What is known of the manner in which Spiders transport themselves from one spot to another by their threads? What is said of the Gossamer Spiders? Describe the architecture of the Labyrinthic Spider. What is said of the Mason Spiders? What of the Spider that weaves leaves together? What of the Spider that builds a raft? What of the Diving-bell Spider? What are some of the Arachnida of the second group? What is said of the Mites? What of the Harvest-men? What of the Book Scorpions?

CHAPTER XXXI.

CRUSTACEANS, AND THE WORM AND LEECH TRIBE.

521. THE class of the Articulata called Crustacea has its name from the Latin word *crusta*, a crust or shell. It includes Lobsters, Crabs, Prawns, Shrimps, Sowbugs, Sand-fleas, Barnacles, etc. Lobsters and Crabs are the most perfect animals of the class.

522. There is considerable resemblance to insects, and also to Spiders, in most of these animals. Like the Insects, they may be divided into two groups—the mandibulate and the haustellate. The eyes of the Crustacea are generally compound, like those of the Insects. They have also antennæ. But the Crustacea differ from insects in the character of their respiratory apparatus. They are aquatic animals, and breathe by gills. There are a few species that are formed to live in air. The Land Crabs, found mostly in the Antilles, are an example. In them there is, above the gills, a spongy apparatus, from which continually exudes a moisture that keeps the gills from becoming dry.

523. The legs of the Crustacea often amount to seven pairs, as in the Woodlouse and Sandhopper; but in other cases there are five pairs, as in the Crab. The legs are constructed very differently in the various Crustacea, according to the manner in which they are to be used. In some they are leaf-like membranes, being thus fitted for swimming; in others they are columns jointed together, to be used only in walking; in others they are so

shaped as to be fitted for digging as well as walking; and in others still they are armed with pincers, so as to be instruments of prehension as well as locomotion. In those Crustacea that swim, as Lobsters, Prawns, etc., the abdomen generally ends in a large fin-like expansion, which works up and down in swimming like the tail of the Whale. But in those which are to walk rather than swim, as the Crab, this part is small, and is bent up underneath.

524. All Crustacea come from eggs. The eggs are commonly carried about adhering to the under part of the abdomen. This we often see in the Lobster. In a boiled Lobster they are red, and the mass is called the coral. More than twelve thousand eggs have been found attached to the abdomen of a single Lobster.

525. There is not generally any true metamorphosis in this class. But in some, the animal, when first born, is entirely unlike the perfect animal. This is the case with the common Crab. In Fig. 240 you see a representation of the Crab when it first issues from the egg. The large figure is a magnified representation, the natural size being given on the little scroll at the side of it. This is almost as unlike the mature Crab as the larva of the Musquito is unlike the Musquito itself (§ 502).

Fig. 240.—Early form of the Crab.

526. In most of the Crustacea there is manifest the ring-like arrangement of segments which is so characteristic of the Articulata (§ 381). But in some it is so much modified as not to be apparent without particular observation. Thus, in the Crab, as we look on its broad carapace of shell, the ring-like arrangement seems to be entirely forsaken; but on examining closely, we find that this carapace is only an excessive enlargement

of one ring encroaching on the others which are still there, although of very small size. We see here the same disposition to have a general plan that we see every where in the structures of nature. A type is always adopted, and we see traces of this in the widest variations from it.

527. The covering of the Crustacea, which is their skeleton, is commonly quite hard, being made so by the carbonate of lime, of which it is in part composed. As this can not grow with the other parts, it must be shed from time to time, and a new and larger covering be formed. The manner in which the old shell is got rid of is very singular. At the proper time there is effected a separation between all parts of the animal and the shell. Then the shell gapes open at some part, and the animal works itself out. This opening, in the case of the Lobster, is down through the middle line of the back. The animal, on emerging, crawls into some by-place where it may be secure, and remains quiet for a day or two till a new shell is formed. The material is supplied from the blood, just as the material for our internal skeleton is supplied from our blood.

528. The Crustacea are divided into fourteen orders. Of these I will notice only a few.

529. The Decapoda, or Ten-footed Crustacea, include the Lobsters, Crabs, Crayfish, Prawns, Shrimps, etc. Nearly all the Crustacea that are used as food are contained in this order. One marked peculiarity of this group is the situation of the eyes on the ends of foot-stalks. The habits of most of these animals are aquatic; but the gills are inclosed in such a way that they do not soon become dry when the animals are in the air, and hence they live for some time after being taken out of the water. They are carnivorous and very voracious; and the first pair of legs are made into powerful claws, by which they seize their food and convey it to the mouth. The mouth itself is quite a complicated appara-

310   NATURAL HISTORY.

tus, there being three pairs of jaws. I have already said enough of the Lobsters and Crabs. The Shrimps and Prawns are quite small animals, regarded as great delicacies. In Fig. 241 the Shrimp is above and the Prawn below.

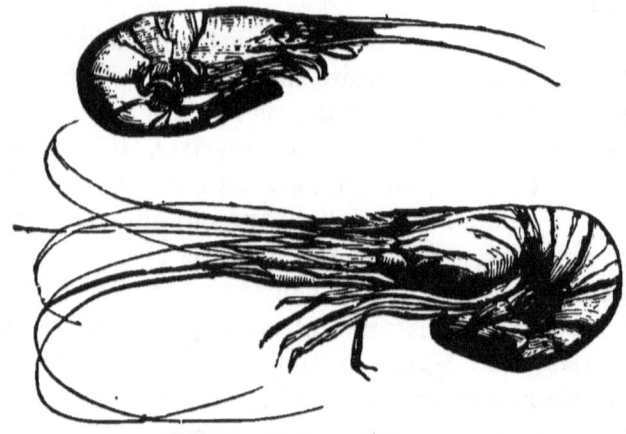

Fig. 241.—Shrimp and Prawn.

530. The Hermit Crabs, Fig. 242, are very peculiar both in their conformation and their habits. The crustaceous covering in the case of these animals is confined to the upper part of the body. The lower part of the body, being uncovered, needs protection, and the animal secures this by inserting its tail into some empty shell which it finds. This it drags about with it as it wanders in search of its food. When it is alarmed, it withdraws itself wholly into its portable house, closing the mouth of the shell with one of its claws. As it grows it is obliged to seek

Fig. 242.—Hermit Crab.

a larger shell, and it is amusing to see one trying one shell after another to find one which will fit.

Fig. 243.—Whale Louse.

531. In the order of Læmodipoda, or jaw-footed Crustacea, is the Whale Louse, Figure 243, which clings by its strong claws to the body of the Whale. So completely is the Whale sometimes covered by these parasites, that a white color is given to its skin, which can be seen at some distance.

532. The order of Cirrhipoda, or tufted-footed Crustacea, contains the Barnacles, Fig. 244, and their allies. The

Fig. 244.—Barnacles.

Barnacle looks like a mussel-shell fixed to a long stem; but, on examination, it is found that the shell consists of five pieces, and through the opening project seven pairs of arms or cirrhi. Two of these are of considerable size, and have suckers on the end, by which they can hold on to any thing. The other six pairs are fringed with cilia, or hair-like filaments, which, by their continual motion, produce currents in the water. This serves both to bring minute animals, constituting the food of the Barnacle, within the reach of the arms, and to move the water over the gills. The animal has jaws which take and masticate the food brought to it by the arms. In Fig. 245 (p. 312) is a Barnacle with the shell partly removed, to show all the parts of the animal. It is always found adhering by the stem to floating wood or

Fig. 245.—Body of the Barnacle.

the hull of a ship. In being inclosed in a shell it is like the Mollusca, and was formerly supposed to belong to that sub-kingdom; but the construction of the animal itself manifestly places it among the Crustaceans.

533. To this order belong also the little Acorn-shells, so called, which are found on the sea-shore in abundance adhering to rocks, shells, etc.

534. The class of Annelida, the Worm and Leech tribe, is one of the lower classes of the Articulata. The animals belonging to it have no articulated members, and there is in them a general inferiority of structure. Still, the lateral symmetry so characteristic of the Articulata, § 387, is retained in them. The two halves of the body are alike. The body is commonly long, slender, and more or less cylindrical. The division into segments, manifest in most of the Articulata, is in this class more manifest internally than externally, it being marked externally only by a wrinkling of the skin.

535. The class is divided into four orders, which I will briefly notice. The first is that of the Dorsi-branchiata (*dorsum*, back, and *branchia*, gill), having the gills arranged in tufts along the length of the body. The animals belonging to this order both crawl and swim with facility. In tropical climates there are some large species, measuring even four feet, and having the body divided into four or five hundred segments. The Sea-centipede, the Sea-mouse, and the Lob-worm belong to this order.

536. The second order is that of the Tubicola, so called because the animals live in tubes. One of the most common is the Serpula, one species of which is represented in Fig. 246 (p. 313). These animals live in shell tubes, attached in groups to stones, shells, and other

THE WORM AND LEECH TRIBE. 313

Fig. 246.—Group of Serpulæ.

bodies. The shell is exuded from the body of the animal just as the covering of a Crustacean is. In the figure one of the animals is stretched up out of its shell, spreading forth its delicate gill-tufts which are arranged around its head. It can withdraw itself entirely within the tube, and when it does so there is a provision for shutting it up. You see that one of the long filaments is expanded at the end into a flat, circular disk. This is the door which shuts down on the mouth of the tube after the other filaments are all drawn in.

537. There are other animals of this group which, instead of having a tubular shell exude from their bodies, form one by connecting together, with a gummy substance from the mouth, particles of shell, sand, small pebbles, etc. They are in this respect like the larvæ of the Caddice-fly, § 459. The Terebella, Fig. 247, does this.

Fig. 247.—Terebella in its Tube.

It is here represented with its tentacula extending out from the tube. These are used in gathering its food. If you take a Terebella, and, breaking up its tube carefully, get the animal in its naked state, you can, by placing it in some moist sand, see the process by which it forms a new tube. In doing this it takes each grain into its mouth, and then, turning its head backward, places it in its proper position.

538. The third order is that of the Terricola, so called because they live in the earth. The Earthworm works through the ground by insinuating its pointed head between the grains of dirt, pushing itself forward by some

little bristly points which all look backward. There are four pairs of them on each segment. It is on account of these that, while you can pass the finger readily on the worm backward, you can feel resistance on attempting to pass it forward. There are two sets of muscles engaged in the movement of the worm—the one longitudinal, which, on contracting, shorten the worm; and the other circular, which make the body smaller and longer when they contract. In Fig. 248 is a representation of an Earthworm at *a*, and at *b* a few segments magnified, so as to show the bristles pointing backward. The egg of the Worm is curiously constructed, having a valve at one end, as seen at *c*. At *d* the young worm has opened the valve, and is coming out. These worms are of great service to the farmer and gardener in loosening the earth below the reach of the spade and the plow. "It has been lately shown," says Carpenter, "that they will even add to the depth of soil, covering barren tracts with a layer of productive mould. Thus, in fields which have been overspread with lime, burned marl, or cinders, these substances are in time covered

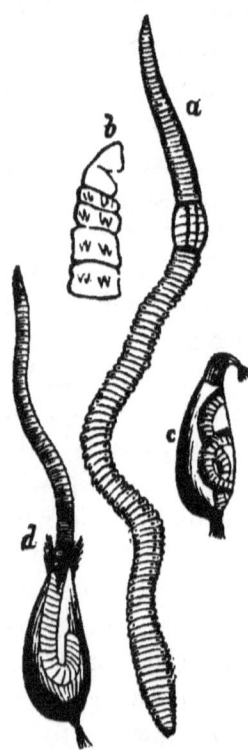

Fig. 248.—Lumbricus Terrestris, or Earthworm.

with finely-divided soil, well adapted to the support of vegetation. That this result—which is commonly attributed to the 'working down' of the materials in question—is really due to the action of the Earthworms, appears from the fact that in the soil thus formed large numbers of 'worm-casts' may be distinguished. These are produced by the digestive process of the worms, which take into their intestinal canal a large quantity of the soil through which they burrow, extract from it the greater part of the decaying vegetable matter it may contain, and reject the rest in a finely-divided state. In this manner a field manured with marl has been covered, in the course of eighty years, with a bed of earth averaging thirteen inches in thickness."

539. The order Suctoria includes the Leech and its allies. The Leech is shaped much like the Earthworm, but has a very different mouth, and a different apparatus of locomotion. It has a sucker at each end of its body, and walks quite fast by fixing the anterior sucker, and then moving the posterior one up to it, and throwing the whole body forward from this. Its mode of walking is much like that of the Measure-worms (§ 479), though its instruments for attachment are different. It can also swim very well by a waving motion of the whole body. Its mouth is in the middle of the cavity of the anterior sucker. In it are three semicircular saws, which make the bite of the Leech. They are so arranged that they work from a central point outward, and make a wound of this $\wedge$ shape. The wound being made, the blood is drawn out by the sucker.

540. The sixth class of the Articulata, that of the Entozoa, includes worms that live in the bodies of various animals, man among the rest. I will notice of this class only those very singular animals which appear to us like long horse's hairs, and are called Hairworms. We see them in stagnant water or in moist places; but they are really inhabitants of the bodies of various insects, and only resort to the water to lay their eggs. If taken from the water and left to dry, they become stiff, horny threads, and appear to have no life; but put them into water again, and they are soon restored to activity.

541. The remaining class, that of the Rotifera, or Wheel Animalcules, contains animals of very minute size, some of them being less than the five hundredth part of an inch in length. Their structure, which is very won-

derful, can only be seen by the microscope; and this, from their transparency, is easily done. They are mostly aquatic animals, and have one or two rows of cilia, or hair-like filaments. It is the motion of these that gives the apparent wheel-like rotation from whence their name is derived.

Questions.—What does the class Crustacea include? What gives them their name? In what respects are they like insects? In what element do most of them live? What peculiar provision is there in one of the exceptions to this? What is said of the legs of the Crustacea? What of their metamorphosis? What of their ring-like arrangement? What of the composition of their covering? What is the necessity of its being shed from time to time? Describe the manner in which this is done in the Lobster. How many orders have the Crustacea? What animals are included in the order Decapoda? What are their peculiarities? What is said of the Shrimps and Prawns? What of the Hermit Crabs? What of the Whale Louse? To what order belong the Barnacles and the Acorn-shells? Describe the construction and habits of the Barnacle. What are the characteristics of the class Annelida? How many orders has it? What is said of the order Dorsi-branchiata? What gives the name to the Tubicola? What is said of the Serpula? What of the Terebella? What is said of the order Terricola? What of the eggs of the Earthworm? What of the usefulness of those animals? What is said of the order Suctoria? What of the class Entozoa? What of the class Rotifera?

CHAPTER XXXII.

MOLLUSKS.

542. THE animals of the sub-kingdom of the Mollusca, or Mollusks are so named from the Latin word *mollis*, soft. Their bodies are soft, and moist, and cold, as you see exemplified in the Oyster and the Slug. All animals that live in shells, with some few exceptions, already noticed, belong to this sub-kingdom. But some belonging to it have no shelly covering, as the Slug and the Cuttlefish, and these are said to be naked.

MOLLUSKS. 317

543. The Mollusks have no skeleton outside or inside. The shells which some of them have are mere coverings, or houses, as we may call them. They do not serve, like the bones of the Vertebrates and the armor of the Articulates, to furnish attachment to the muscles so that they may act. Those Mollusks that lead the stillest life, that is, which use their muscles least, generally have the firmest and thickest shells.

544. The shell is composed of *carbonate* of lime, with some animal matter, while in the bones of the Vertebrates the mineral portion is the *phosphate* of lime. In some the mineral part predominates, and the shell is very hard, like porcelain; while in others, as the oyster, there are distinct layers of the mineral matter, with a membrane of animal substance between them. The shell is secreted from the thick skin of the animal, which is called the *mantle*. It is formed from the blood, and the materials for it are taken in with the food.

545. Shells are of two kinds—those which are in one piece, and those which are in two pieces, with a hinge to keep them together. Mollusks that have the first kind of shell are termed *univalve*, and those which have the second are termed *bivalve*. Clams and Oysters are familiar examples of bivalves. Two varieties of univalves are represented in Fig. 249.

Fig. 249.

546. Shells undergo some changes in form as they grow

with the growth of the animals in them. Sometimes additions are made to them, entirely altering the figure, so that two animals of different ages really of the same species would hardly be recognized as such. In Fig. 250 we have at *a* and *b* back and front views of the shell of a

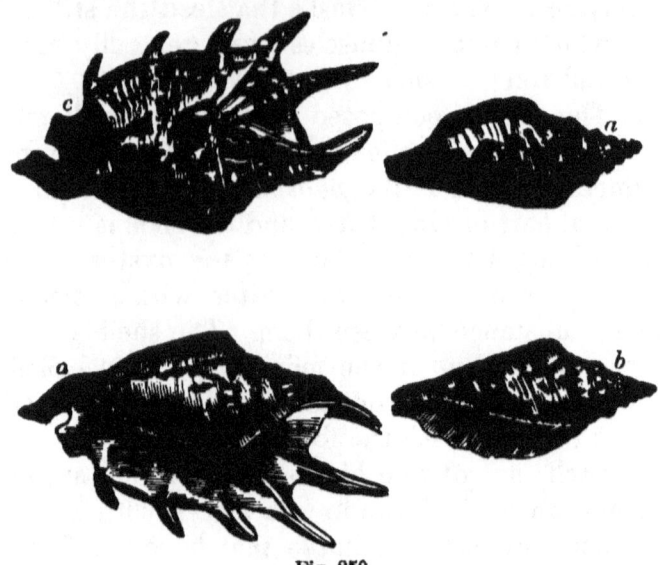

Fig. 250.

young Mollusk, and at *c* and *d* similar views of the shell of the full-grown animal. The addition of the spines bears some analogy to the addition of horns in some of the Mammalia.

547. Most of the Mollusca can move about but little, and some none at all. They have but little muscle, and are in this respect in striking contrast with the Articulata, which are nearly all muscle (§ 383). It is only where the body is naked (that is, without a shelly covering), or where a portion of the body can be projected out from the opening in the shell, that any active movements can be effected. In many inhabiting *bivalve* shells there is a fleshy, tongue-like projection called a *foot*, which in some cases enables the animal to leap; in some is used as a boring apparatus; in some acts as a sort of fin for swim-

ming; and in some produces the *byssus*, a collection of threads by which the animal attaches itself to rocks and other objects. In most of those which inhabit *univalve* shells there is no projecting foot; but the under side of the mantle is thickened into a fleshy disk, which by its contractions and expansions effects the progression of the animal, as is seen in the common Snail. Among the Mollusks similar to these in structure, but having no shell, the whole mantle is muscular, enabling them to move quite freely, especially those that live in water. In the Cuttle-fish tribe we have the most efficient means of locomotion in the shape of arms, and in some of this group there are fin-like appendages, the arms being quite short.

548. Leading such a sluggish life as most of the Mollusks do, their destiny seems to be to grow, by their digestive powers, into a well-fatted mass, so that they may be good food for other animals that inhabit the deep, and some of them for man.

549. Almost all of these animals breathe by gills; but some, like the Snails and Slugs, have something like lungs, as they live in air. The blood is nearly colorless, and circulates in a regular system of arteries and veins connected with a heart.

550. This sub-kingdom has two grand divisions—the Cephalous Mollusca (κεφαλή, *kephale*, head), those which have heads; and the Acephalous, those which are headless. I will first speak of the Cephalous. All belonging to this division that have shells have those which are univalve. The Cephalous Mollusks are divided into three groups: 1. Cephalopoda, those which have feet arranged in a circular manner around the head. 2. Pteropoda, wing-footed. These have a pair of wing-like expansions of the mantle, which serve as fins, and enable them to swim quite rapidly. This is a small class, but a very interesting one. 3. Gasteropoda, belly-footed. These have a single broad *foot* on the under surface of the body. The first two classes belong entirely to the sea; but this

class has some species that live in fresh water, and some even that live on land. I will notice each of these groups.

551. Of the Cephalopoda, the only few existing species that have a shelly covering are the Argonauts and the Pearly Nautilus. There are, however, many fossil shells found which must have belonged to animals of this group. The Ammonites, commonly called Snake Stones, of which a specimen is given in Fig. 251, are the most abundant of these, there having been described over five hundred species. These are found in various kinds of rocks, and are of various sizes, some reaching a diameter of even four feet.

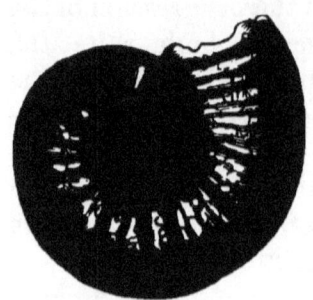

Fig. 251.—Ammonite.

552. The arms of some of the Cephalopods are very long. This is the case with the Cuttle-fish, one of the most singular of animals, seen in Fig. 252. Its body is

Fig. 252.—Cuttle-fish.

soft, and is covered only with a leathery skin. From around its mouth extend eight long arms, which have on them great numbers of little suckers, by which it can

MOLLUSKS. 321

cling to rocks or retain its hold upon its prey. It has a powerful parrot-like beak, with which it can crush the shell-fish and the crustacea that it captures. It can manage even a large Crab in this way. Winding its long arms around it, and holding it, both body and claws, with its numerous suckers, it deliberately crushes its various parts with its strong mandibles, and picks out the flesh In the Indian seas this animal attains so large a size as to be a dangerous enemy even to man. The color called sepia comes from the Cuttle-fish. It is used by the animal for darkening the water with an inky cloud, that it may more easily escape from a pursuing enemy. The so-called cuttle-fish bone is a chalky substance secreted from the mouth of the fish, and is composed of almost innumerable plates united by myriads of little pillars.

553. The Argonaut, Fig. 253, called the Paper Nautilus, from its thin, white, delicate shell, has, like the Cuttle fish, eight arms with suckers. Two of these are expanded into broad membranous flaps. From early times it has been said that this animal uses its membranous arms for sails, and its other arms for oars It has been found, however, that the membranes are not used at all as sails, but are usually spread over the sides of the shell, meeting along its keel. It is from them, and not from the surface of the body, that the calcareous secretion is poured forth for the enlarge-

Fig. 253.—Argonaut, or Paper Nautilus.

ment or reparation of the shell. It is by the action of the arms as oars, and by the forcing out of water from the gill-chambers, that the animal can swim. Both the Argonaut and the Cuttle-fish use their arms as feet to walk along on the bottom of the sea.

554. The Pearly Nautilus is so called from the "nacre," or mother-of-pearl with which its shell is lined. It is found on most shores between the tropics. It is peculiar in having many separate chambers in its shell, in only one of which, the largest and the outermost, the animal lives. It has a connection, however, with the other chambers by a membranous tube called the siphuncle. It is supposed that the animal, when it wishes to sink in the water, can force some water into this siphuncle, thus increasing its specific gravity; and that the reverse takes place when it wishes to rise. Some doubt this, and consider the design of the siphuncle and the chambered structure as yet a mystery.

555. The Pteropoda, or wing-footed Mollusks, constitute a small and aberrant group. The animals included in it may be considered as having the same place in the Molluscous kingdom that the Birds have in the Vertebrate and the Insects in the Articulate. They fly in the water, having for the purpose a pair of fin-like organs, or wings, which are an expansion of the mantle on each side of the neck. Though the number of species in this group is small, the number of individuals in some of the species is often enormous in some localities. Some have a shell and some have not.

556. The Clio Borealis, Fig. 254, one of the best known of this class of Mollusks, is very abundant in the arctic seas, and is one of the principal articles of food of the Whalebone Whales (§ 192). These little animals are sometimes so numerous that the Whale can not open its mouth without ingulfing thousands of them. The Clio has eyes, which, though exceedingly small, are very perfect in their organization. It has powerful jaws armed

Fig. 254 —Clio Borealis.

with teeth, calculated to tear in pieces the minute animals on which it feeds. It has also a very effective apparatus for securing its prey, consisting of six tentacula of a reddish color. On examining one of these with a microscope, this color is found to be occasioned by red points arranged with great regularity. On magnifying these still farther, each point is seen to be a collection of about twenty suckers on the ends of as many stalks. Each collection is in a sort of sheath, and can be protruded from it. There are on all the tentacula about three hundred and sixty thousand of these suckers, constituting an apparatus for prehension more extensive, in proportion to the size of the animal, than any other to be found in the whole animal kingdom.

Questions.—What is the significance of the name of the third sub-kingdom of animals? What are naked Mollusks? What is said of the use of the covering which most of them have? Of what is it composed? What is said of the proportions of the constituents? How is the shell formed? What are the two kinds of shells? What is said of the changes which shells undergo in growing? What is said of the locomotion of Mollusks? What is said of the foot, and its various uses? What is the byssus? What provision for locomotion is there in most of the Mollusks that inhabit bivalve shells? What in those that are similar in structure, but have no shell? What is the special destiny of Mollusks? What is said of their breathing apparatus? What of their blood, and its circulation? What are the two grand divisions of Mollusks? What are the groups in the first division, and their characteristics? Of the Cephalopods, what shelly species exist at the present time? What is said of the Ammonites? Describe the structure and habits of the Cuttle-fish. What is sepia? What is Cuttle-fish bone? What is said of the Argonaut? What of the Pearly Nautilus? What is said of the Pteropod group? What of the Clio Borealis?

CHAPTER XXXIII.

MOLLUSKS — *continued*.

557. The class of Gasteropoda is mostly composed of Mollusks that live in a *univalve* shell, which is usually of a spiral shape. You have two different forms of the spiral in Fig. 249, page 317. Some of the species, as the Slug, are naked or destitute of shell. There is, however, in these, sometimes a small shell, generally imbedded in the mantle, just over the cavity which contains the lungs. The body of the Gasteropods is terminated in front with more or less of a head, having fleshy tentacula, varying from two to six in number. The back is covered with a mantle which secretes the shell. On the under side of the animal is the fleshy mass called the foot. In those which have a shell, all the body remains in it except the head and the foot. These project beyond it when the animal expands them for walking, but they can be withdrawn into the first turns of the shell at pleasure. In most of the aquatic Gasteropods there is on the foot a plate of horny substance, which shuts over the opening

Fig. 255.—Limnæa Stagnalis.

in the shell after the head and the foot are drawn in. In Fig. 255 you see one of these animals with the head and the foot out of the shell.

558. Many of the Gasteropods are remarkable for an abundant supply of flinty teeth. Sometimes these are on the palate, and in some species even the stomach has teeth scattered over its inner surface. The tongue, in some, is remarkable for its length, and for the teeth which are all along on its upper surface. The tongue of the common Limpet, Fig. 256, is an example. It is from two to three inches long, and this is longer than the whole animal. When not in use, it is turned backward down into the stomach. It is spoon-shaped at the end. In its whole extent it is armed with rows of teeth, four in each row, and between each two rows there are two three-pointed teeth. These two sets of teeth are represented in a magnified portion of the tongue in the figure. The part of the tongue toward its root generally has its edges turned over so as to meet, thus making a tube. The whole instrument is therefore an efficient rasper, and also a proboscis.

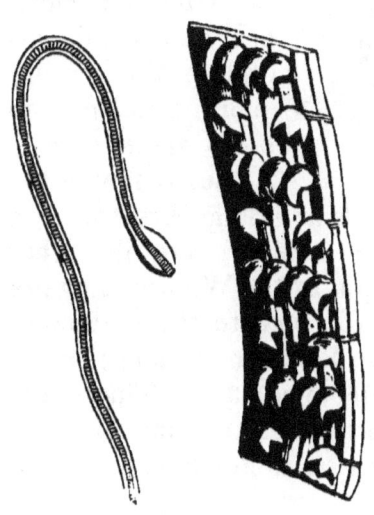

Fig. 256.—Limpet's Tongue.

559. Of the Gasteropoda, some are terrestrial and some live in fresh water, but most of them are found in the sea. The terrestrial Gasteropods are Snails and Slugs. In the common Slug there is a prominent head with four tentacula, which can be drawn inward by a process like the inversion of the finger of a glove. At the ends of the longer pair of the tentacula are the eyes. On the back there is a kind of shield formed by the mantle,

which usually incloses a small shell. This shield is over the breathing apparatus (§ 557), and the head can be so drawn in as to be under it. The Snails have very much the same shape and arrangement with the Slug, except that they have a shell into which they can withdraw the whole body. The common Snail, Fig. 257, lays eggs, which are very large in comparison with the size of the animal. They are of the size of a small pea, and are deposited in the ground about two inches below the surface.

Fig. 257.—Snail.

560. A few of the Gasteropods that, like the Snails and Slugs, breathe with lungs, are yet aquatic in their habits. But, like other aquatic animals that have lungs, as the Whales, they are obliged every now and then to come to the surface to get air. Among these are the Pond Snails, a species of which is represented in Fig. 255, page 324. These Mollusks, and those which are terrestrial, the Slugs and the Snails, are included in an order by themselves, as having lungs—the order **Pulmonifera**.

561. The second order of the Gasteropods includes all those which have gills instead of lungs, and also have a shell, usually of a spiral form. This order is much larger than the others, and presents a great variety of beautiful shell-coverings. Some of them have siphons to introduce water into the cavities where the gills are, so that the animal can breathe without putting its body out from the shell. There is a little notch always to be observed in the shell where this siphon passes out.

562. Of the many varieties of the shells of these Gasteropods I will notice but a few. In Fig. 249, page 317, on the left, is an example of the Turbinidæ, or Whorl fam-

ily, called the Royal Staircase Wentletrap. This is found in the Chinese and Indian Seas. It is so costly—a fine specimen commanding, even now, four or five pounds sterling—that the specific name attached to it is *pretiosa*, precious. In the same figure is a specimen from the very extensive Cone family. In Fig. 258 the large shell is that

Fig. 258.

of a Whelk, belonging to a family which, from the shape of the shells, is called Buccinidæ, from *buccinum*, a trumpet. The famous Tyrian purple was obtained from one of this family. In the same figure is the little Cowry, which is a current coin among the natives of Bengal, Siam, and many parts of Africa. In Bengal, 3200 of these shells are reckoned equal to a rupee, or about two shillings of English money. In 1849 about three hundred tons of them were imported into Liverpool, designed to be used in the African trade. One of the most beautiful of the shells which are armed with spines is the Thorny Woodcock, Fig. 259, sometimes called Venus' Comb.

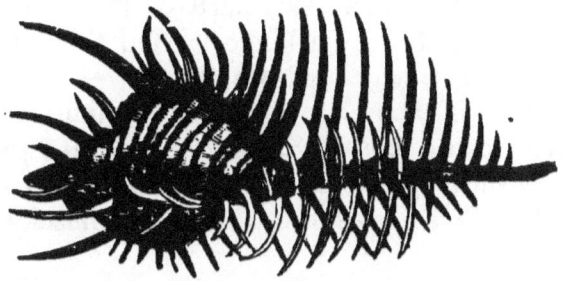

Fig. 259.—Thorny Woodcock.

563. There is a third order of the Gasteropods, in which the gills are not in a covered cavity or chamber, as they are in the second order, but they either stand out on the back, or are more or less concealed at the sides in folds of the mantle. Some of them have shells, but most have not. I will give but a single example, the Glaucus, Fig. 260, found in the Mediterranean and Indian Seas.

Fig. 260.—Glaucus Atlanticus.

The hues of these beautiful animals are azure blue and silver. The gills form two or three large tufts on each side, which, besides being the breathing apparatus of the animal, are also its instruments for swimming.

564. We now come to the second grand division of the Mollusks—the Acephalous or Headless Mollusks. These may be divided into two groups: 1. Those which have shells, called the Conchiferous, or shell-bearing. 2. Those which are covered with a leathery or membranous tunic, called the Tunicated.

565. The shells of almost all the Conchifera are bivalve. This group includes the Oysters, Clams, Mussels, Scallops, etc. The shell is exuded or secreted from the mantle, and is in different layers, as may be seen in the shell of the Oyster. The outermost layer is the smallest, and as the animal grows, each layer is a little larger than the one outside of it. The two parts or valves of the shell are joined together by a hinge. Near this hinge is an elastic ligament, which allows the valves to be a little apart, which is their natural position, admitting the water freely to the mouth of the gills. When the animal wishes to shut the valves closely, it does so by means of a muscle. Sometimes there are two muscles for this purpose.

566. That you may understand the plan of the organs

MOLLUSKS. 329

of these animals, I will give you the anatomy of one of
them in Fig. 261. One of the valves is removed. You

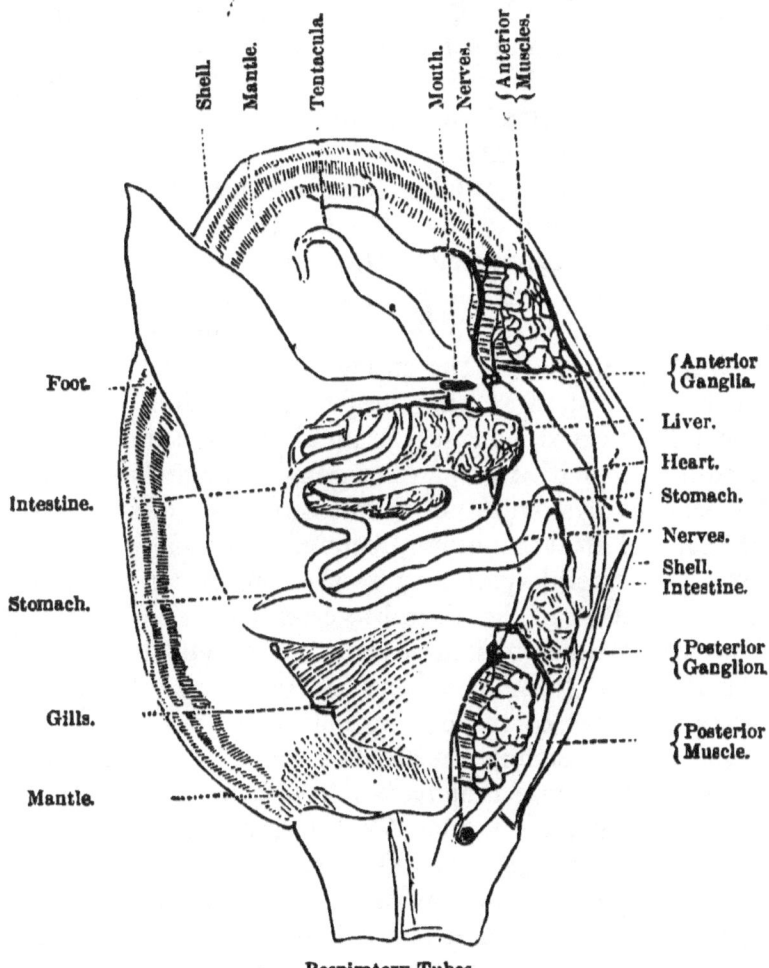

Fig. 261.—Anatomy of an Acephalous Mollusk.

see the mantle, fringed all around its edge. This lines
the whole shell, and covers the animal. It is its skin.
You see the two muscles that, by their contraction, bring
the valves together, and the fleshy foot, which can be
made to protrude when the valves are left to go apart by
the action of the elastic ligament. This foot, which is

the only locomotive organ that the animal has, serves, in different species, a variety of purposes, sometimes enabling the animal to leap, sometimes being used to bore into sand or mud, and sometimes only serving to fix the animal to some solid support. In some there proceed from this foot a band of hair-like filaments, called the *byssus*. While fastened to some object by these filaments, the animal may have some considerable motion within certain limits. The gills have two respiratory or breathing tubes connected with them, by one of which the water passes into the gills, and by the other passes out. The water is made thus to go in and out by fine cilia in the gills and on the surface of these tubes, which keep up a constant waving or fanning motion. There are certain nerves, you see, branching about, and they are connected with two pairs of ganglia, or little brains. The nervous system is very limited, for the animal has little need of either thinking, feeling, or motion.

567. The lateral symmetry, so thoroughly observed in the construction of the Vertebrates and the Articulates, which was forsaken to some extent in the Cephalous Mollusks, is in the Acephalous entirely given up. In them there are no two corresponding halves of the body.

568. The Conchifera we divide into two sections—the first including those that have not siphons, and the second those that have them. To the first section belong the Oysters, Scallops, Pearl Oysters, etc. The shell of the Oyster has two unequal valves. One of these bulges out more than the other, and it is by this that it is fastened to rocks, or pieces of wood, or to other Oysters. The structure of this animal is even more simple than that sketched in Fig. 261. It has no foot; for, as it is fixed by its shell in one spot, it needs none. Oysters are very prolific animals, forming immense *banks*, and thus providing quite largely for the sustenance of man. "But man," says Carpenter, "is by no means the only enemy to the Oyster. Its body serves as food to many marine

animals, which have various modes of getting at it, in spite of its shelly defense. From some of these it can secure itself by closing its valves as soon as it is alarmed; and against others it has a more active means of defense in the violent expulsion of the water included between them, which (as it is itself fixed) will frequently drive off its opponent. Various animals attack it, also, by perforating its shell; and to these, also, it can offer a passive resistance, by depositing new shelly matter within. So that even this lowly-organized being, commonly regarded as one of the most vegetative of animals, is provided by its Creator with such means as are necessary for its preservation, and doubtless, also, for its enjoyment."

569. Pearl Oysters, from which pearls are obtained, are found both in the Old and New World. Ceylon is famous for its pearl fisheries. Pearls are globules of "nacre," which chances to be deposited in this form, instead of being spread out over the inner surface of the shell; it being in the latter case called mother-of-pearl. The Pearl Oyster is not the only animal from which pearls can be obtained. They are often found in other shells.

570. The Pectens, or Scallops, of which a species is given in Figure 262, are distinguished by the regular ribs of the shell, and by the two angular projections that widen the sides of the hinge. They have a small foot, and some species have a byssus. In some the shell is beautifully colored.

Fig. 262.—Scallop.

571. Among those Conchiferous Acephala that have siphons are the Clam-shells, the Cockles, etc. Among the Clam-shells is one which is the largest known Mollusk. It is the Tridacna, or Giant Clam-shell, found only in the In-

dian and Australian waters. There is a pair of these shells in the Church of St. Sulpice, in Paris, used as receptacles of holy water, which weigh over five hundred pounds. The common Clam belongs to a different group, the Veneraceæ. The foot of these, of the Cockles, and of the Pholadaceæ, the group to which the Teredo belongs, is used mostly for burrowing. Most of them burrow in sand or mud, some in rocks, and some in wood. Those that burrow or bore in hard substances can not do this with the foot. It is done with a sort of rasping operation of the edges of the shell, the foot answering in this case only as a means of holding on while the animal bores. The Teredo, by this boring operation, is largely destructive to ship bottoms, piles, etc. Holland has been sometimes threatened with an inundation by the destruction of dikes by this little Mollusk.

572. One of the most interesting of the Mollusks which burrow in sand is the Razor-shell, so called from its length. It can burrow very rapidly, and therefore it is quite difficult to catch it. It bores in the sand with its foot, which it can elongate so as to make it quite pointed. Its burrow is recognized by the little jet of water which it throws out when it is alarmed. If a little salt be thrown upon its hole it will make its appearance, but one must be quick in his movements to catch it before it can get out of sight again. Its mode of burrowing is very curious. It puts its foot into a dagger-shape, as represented at *a* in Fig. 263, and thrusts it downward in the sand. Now it gives it the shape of a bell-clapper, as at *b*, and the end furnishing it a hold in the sand, it moves its body forward by shortening the foot. By repeated movements of this kind it gets along quite rapidly in the loose sand.

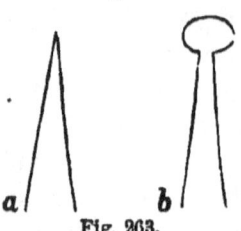
Fig. 263.

573. What is stated in the previous paragraph exemplifies one of the many modes in which the foot of Mol-

lusks is used. Some, thrusting it out, attach it to some support, and then, by contracting it, pull themselves along. Some use it to push themselves forward, as a man in a boat pushes himself from the shore with his oar. And some, by bending the foot and then quickly straightening it, leap forward. There is a little Mollusk, the Ianthina, or Oceanic Snail, Fig. 264, which has attached

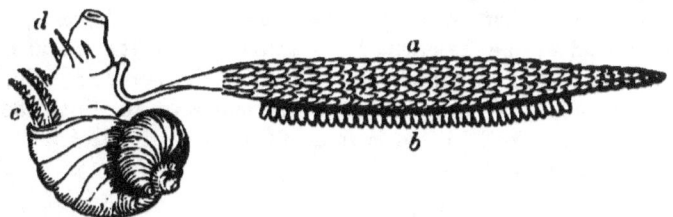

Fig. 264.—Ianthina with its raft.

to its foot a raft of singular construction. It is made of numberless vesicles, *a*, filled with air. Its purpose is to float the eggs, *b*. You see at *c* the gills of this little animal, and at *d* its tentacles and eye-stalks. The Ianthina is often met with in great numbers in companies in the open sea. In rough weather much damage is often done to their beautiful floats, and sometimes they are wholly destroyed.

574. The Tunicata form an aberrant group of the Mollusca, verging, in their organization, toward the Radiata, the only remaining sub-kingdom to be noticed. Although it would be interesting to consider this group, I shall pass it by. I shall also omit another class, the Polyzoa, formerly supposed to belong to the Radiates, but recently ascertained to belong to the Mollusks.

Questions.—In what do the Gasteropods mostly live? What is said of the form of their shells? What is said of the naked Gasteropods? Describe the structure of these animals. What is said of their teeth? Describe the structure of the Limpet's tongue. What are the terrestrial Gasteropods? Describe the common Slug. Describe the common Snail. What is said of the Pond Snails? What are included in the order of Gasteropods called Palmonifera? What is said of the order whose animals breathe by gills? What is said of the Whorl family? What of the family called Buccinidæ? What is said of the Cowry? What is said of the third order of Gasteropods? What are the two groups of Acephalous Mollusks? What are the shells of the Conchiferous group? What does it include? In what way is the shell formed? How are the two valves united? How moved? What is the anatomy of the Acephala? What is said of the symmetry of these animals? What are the two sections of the Conchifera? What is said of the Oyster? What of the Pearl Oysters? What of the Pectens? What are among the Conchifera that have siphons? What is said of the Tridacna? What is said of the Cockles, the Veneraceæ, and the Pholadaceæ? What of the Teredo? What of the Razor-shell? What is said of the various ways in which the foot is used by Mollusks? What is said of the Ocean Snails? What is said of the Tunicata?

CHAPTER XXXIV.

RADIATES.

575. WE now come to the last sub-kingdom—that of the Radiates. The arrangement of structure here is, in many respects, entirely different from that of the other sub-kingdoms. There is a lateral symmetry of form in the Vertebrates and the Articulates. While this is mostly abandoned in the Mollusks, in the Radiates it is exchanged for another symmetry of a wholly different character — a symmetry of rays arranged circularly. It is therefore akin to that of plants. Indeed, many of the animals of this sub-kingdom were formerly supposed to be plants, and are now, from the resemblance referred to, called plant-animals.

576. This resemblance may be very distinctly seen in the Actiniæ, or Sea Anemones, of which there are many species. The structure of these is very singular. There is a broad, flat, muscular base, of a circular shape, by which the animal adheres firmly to a rock. From this base rises a rounded body, on the top of which there is an orifice, which is more or less open according to cir-

cumstances. The animal can close this opening entirely. Around this mouth of the animal are arranged rows of tentacula, extending out like rays when the mouth is open, and giving the creature the appearance of a flower with its spread petals. Fig. 265 shows Sea Anemones in

Fig. 265.—Actiniæ, or Sea Anemones.

three different states. The upper one has the mouth closed and the tentacles drawn in, and the animal presents almost a hemispherical form. The one just below on the rock is partly opened; and another, under the water, is fully expanded, looking like a flower.

577. The mouth of the Actinia conducts to a stomach which may be said to be very large in proportion to the whole body. The office of the tentacles is to catch the prey of the animal, and force it into this cavity. Fig. 266 represents one of these animals cut open, showing the stomach at a. At b are certain cavities or chambers, which are all around the stomach. These chambers all communicate with each other, and also with the tentacles, which are tubular.

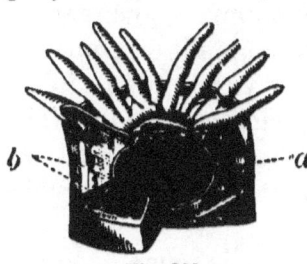

Fig. 266.

Water is taken into the chambers by these tubes, and then is forced out, through these same tubes, in jets, with such force, often, as to rise to the height of a foot or more. The chief office of these chambers seems to be to expose the blood of the animal to the air in the water. In other words, they are the gills, or the breathing apparatus. These animals are found on all coasts, commonly on rocks, where they can be a part of the time under the water and a part of the time out of it. Their habits I shall refer to again hereafter.

578. The radiate arrangement so manifest in the Sea Anemones and in the Starfish (§ 17) is not seen so plainly in many of the other animals of this sub-kingdom; and some of the orders are quite aberrant. In some there is a considerable approach to the Articulata. The Starfishes and the Sea Anemones are among the type families of the Radiates.

579. Some of the animals of this sub-kingdom have the power of moving about, but most of them, in conformity with their plant-like character, are stationary during a part or the whole of their existence. In muscular apparatus, therefore, most of the Radiates, like the Mollusks, are in strong contrast with the Articulates.

580. There is a resemblance to vegetables in still other respects besides those already mentioned. When any parts of these animals are lost by accident, they are generally replaced by a new growth. Besides, there is often a new animal produced entire by a sort of budding from some part. And even farther than this, in some portions of this kingdom the animals are arranged in companies, like the parts of a plant, on a common stalk or trunk.

581. None of these animals have any thing like a head, and they have only the senses of touch and taste. The senses of sight, hearing, and smell, so far as we can see, are wholly absent. For the arrangement of the nervous system, I refer you to § 18.

582. As in most of the Radiates there is a small amount

of muscle, there is a very remarkable structure which seems in some respects to take its place. This structure, though found to a considerable extent in other animals, is present to an extraordinary degree in the Radiates. It is the ciliary structure alluded to in § 566. Cilia are fine hair-like filaments which cover the surface of many membranes, and fringe their edges. They are quite regularly arranged, sometimes in straight rows, and sometimes spirally or in circles. They have a motion which, in some cases, is obedient to the will of the animal, but in others is independent of the will. When in motion each filament bends from the root to its point, straightening out again, like a stalk of grain acted upon by the wind; and we have, therefore, when many of them are in motion, an appearance like the successive waves in a field of grain as the wind blows over it. This motion can be seen only by the aid of the microscope. It is beautifully displayed in the gills of the Oyster. The object of this movement is to produce currents in the fluid in contact with the membrane. These currents serve various purposes, as, for example, to bring food within the reach of the tentacles, and to carry fresh portions of water through the respiratory apparatus. For this latter purpose cilia cover the membranes lining the chambers in the Actiniæ (§ 577). Cilia are needed in those animals which are most stationary, and in them, therefore, they are most manifest.

583. We divide this sub-kingdom into three classes: 1. Echino-dermata (εχῖνος, *echinos*, a sea-urchin; δερμα, *derma*, skin), prickle-skinned animals. 2. Acalephs (ακαλήφη, *akalephē*, a nettle), Sea-nettles, or Jelly-fishes. 3. Phytozoa (φῦτον, *phyton*, a plant; ζῶον, *zōon*, an animal), commonly called Polyps. These are fixed, like plants, and have flexible arms about the mouth, as seen in the Sea Anemone, Fig. 266.

584. One of the Echinoderms, the Starfish, I noticed in the first chapter (§ 17). It merits here, however. a more particular description. It is only the upper side of

the animal which is represented in Fig. 9. On its under side are great numbers of little feet. With these it walks along on the bottom of the sea, searching for food, which it puts into its mouth, this being in the centre of the star on the under side. These feet are fleshy, and are hollow tubes, like the tentacles of the Actiniæ (§ 577). They are so shaped that they can be used as suckers, and the animal can shorten and lengthen them at pleasure. It is by pumping water into and out of them that the suction is effected. In walking, the suckers are some of them thrown forward, and, taking hold of the surface on which the animal is, and then shortening, they draw it forward. It can walk up the side of a smooth rock in this way. The operation can be seen by placing one of these animals in a glass vessel filled with water. If you place a Starfish in your hand on its back, that is, with its feet upward, you will see these little suckers reaching forth in all directions; and if you look at them with a magnifying glass, you will observe a ring-like arrangement in each sucker as it lengthens out, quite as plainly as you see it in a common worm.

585. These animals not only walk with these suckers, but they seize their prey with them. They are carnivorous and rapacious; and in taking their prey they fasten their suckers to it, and work it up to the central mouth, which is opened wide to receive it.

586. Besides the motion of the suckers, the five arms on which these are can be moved also in various directions. In some species there are little red spots at the ends of the rays, which are supposed by some to be eyes; but this is very doubtful.

587. The order Stellerida, to which the Starfish belongs, includes a large variety of animals having a general resemblance, but varying in the relative proportion of the body and the rays, and the arrangements of the latter. In some species there is little else but arms, while in others the central part is large.

588. The Sea Egg, as it is commonly called, is the crust or shell of a spiny or prickle-skinned animal, stripped of its spines. In Fig. 267 you see this animal, called an

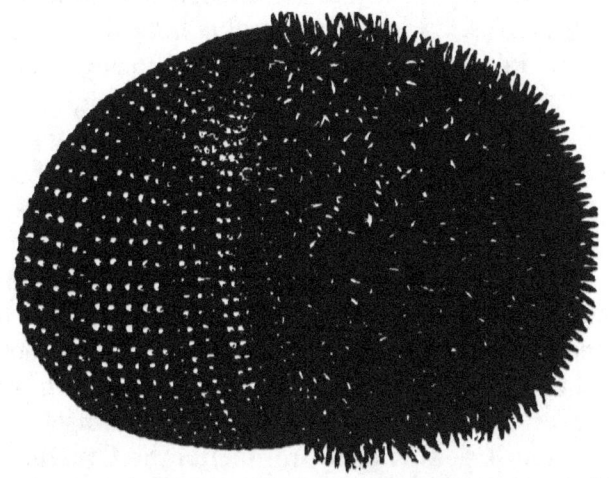

Fig. 267.—Shell of Echinus, or Sea Urchin; on the right side covered with spines, on the left the spines removed.

Echinus, with the spines removed from half of it. These spines are curiously jointed with the shell. There is a round projection of the shell at the root of each spine, upon which the spine works with its cup-like cavity, making a regular ball and socket joint. These projections every one must have noticed arranged with such beautiful regularity on the Sea Egg. There are the same tubular feet as in the Starfish, but much larger, and therefore more efficient in taking prey. In walking, while the suckers are the moving power, the animal is carried forward on the spines, these acting after the manner of a crutch. The animal inside of this singular shell has a stomach, a respiratory apparatus, intestines, etc. Its mouth has quite formidable teeth. Small Crustacea and Mollusca are its chief food.

589. The shell is made up of small plates, and, as the animal grows, each one of these plates is made larger by increase at its edge. The growth is like that of the cov-

ering of the Turtle. If it were not for this arrangement the animal would be obliged to leave its shell occasionally, and have, like the Lobster, a new covering formed.

590. The Echini (plural of Echinus) are generally found on sandy shores. Here they make hollows with their spines, and in them lie in wait for their prey. As they do this they let their tubular feet play about, and when any Mollusk or Crustacean happens to hit a sucker, it is at once captured, many suckers taking hold of it, and passing it to the mouth to be crushed, and thrust into the stomach.

591. Many of these animals have a powerful and complex masticating apparatus. It consists of five hard, sharp teeth, worked by strong muscles. These teeth are attached to bony jaws, and the whole apparatus has twenty-five pieces, moved by thirty-five distinct muscles. It is a powerful mill, reducing to fragments the Crustacea and Mollusks which the tentacula capture and force into it.

592. The most singular of all the facts in regard to the Echini is the mode of their development. There comes out of the egg an animal covered with cilia, and by the waving movement of these, it swims freely about in the water. At first it is globular, but it soon acquires a pyramidal form, having a stomach opening below. At the same time there are formed four slender, bony rods in the four angles of the pyramid, meeting together at the top. There are some cross-pieces, also, on the sides of the pyramid, connecting the rods together. All this time the animal is moving about by means of the cilia, which are all over its outside. It is a sort of pyramidal tent sailing about. Inside of this the real animal is at length formed, and, at the same time, the tent-portion of the original animal wastes away. The stomach of the animal that comes out of the egg is the only part which remains through all this metamorphosis.

593. There are two orders of the Echinoderms which are quite aberrant. One is that of the Crinoidea, which

derives its name from the lily-like form which some of its species present. Most of the species are extinct, but they are found in their fossil state abundantly in limestone and some other rocks. The other aberrant order is that in which are those animals that are called by sailors Sea Cucumbers, from their resemblance in form and in surface to the cucumber of our gardens.

594. The second class of the Radiata is that of the Acalephs. These animals are called Sea Nettles and Stangfishes, from the stinging sensation which nearly all of them can inflict on being touched. They are also called Jelly-fishes, from their great softness. Most of their bulk is merely water. Though one may weigh even many pounds when first taken from the water, when it has lost all its fluid parts it will weigh only as many grains. There are many species, some being no larger than the head of a pin, and some being of very considerable size.

595. One of the most common of these animals is the Medusa. This is often seen in great multitudes floating along near the shore in a calm, bright day. You see the shape and usual position of the animal in Fig. 268, B.

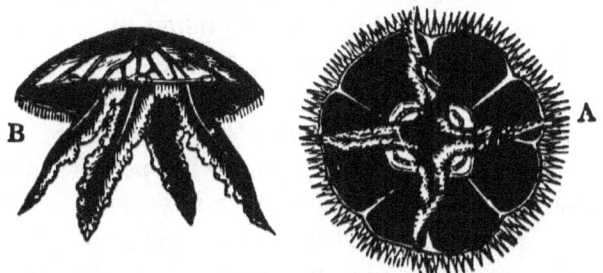

Fig. 268.

Its body is umbrella-shaped, with a fringe around its edge. It is by a waving motion of this umbrella that it moves along in the water. Its mouth is in the centre of the under surface, and from around it hang down four leaf-like tentacula, which are both feelers and graspers of its prey. These tentacula carry the food to the mouth. The stinging power possessed by them is probably of service in overcoming its prey, like the poison of the Scorpions and other insects. At A you see the under surface, showing the mouth in the middle. The resemblance in arrangement to the Actiniæ is very obvious, the chief difference being that, in the one group, the mouth is above, while in the other it is on the under surface. The Medusæ often reach a considerable size. It is said that they have been seen of three or four feet in diameter, and of even sixty pounds weight. Although they are such watery animals, they eat solid food, for in their stomachs have been found small Crustacea, Mollusks, and even Fishes.

596. The Acalephs generally float near the surface of the water, and sometimes are seen in great abundance basking in the sun, and reflecting its rays in such a manner as to make a play of the most brilliant colors. The phosphorescence sometimes seen in the sea is owing chiefly to small Acalephæ. Carpenter thus describes the beauty of this phenomenon as witnessed in the warmer latitudes: "The whole surface of the ocean displays a diffused luminosity, like that of the Milky Way on a clear night. The path of the ship is marked by a brilliant line of glowing light. The waves, as they gently curl over one another (this phenomenon is never seen with a *rough* sea), break into brilliant spangles. The oars of a boat rowing over them seem dripping with pearls when raised from the water, and every stroke is marked with a new line of brightness. And amid this general splendor, varied forms of more glowing lustre are seen to move—some like ribbons of flame, some like globes of fire, some gently gliding through the still ocean, others more rapidly moving just beneath its surface."

597. To the Ciliograde order of Acalephs belongs the common Beroe, Fig. 269, which is thus described by Dr. Harvey, an English naturalist: "This little creature is

RADIATES. 343

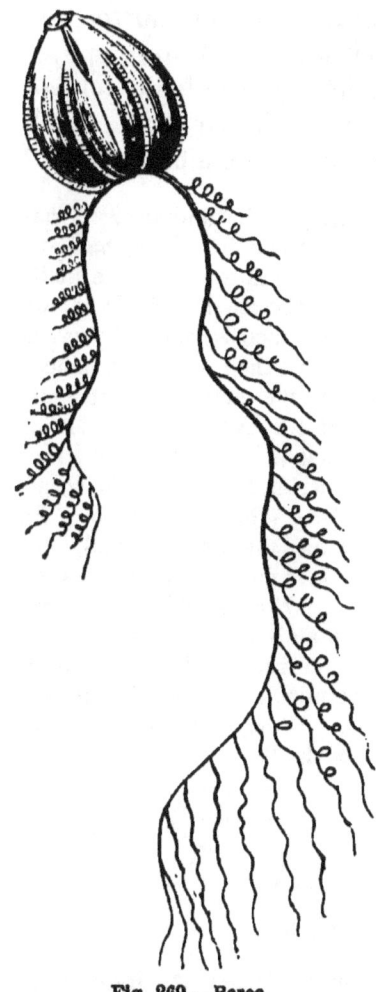

Fig. 269.—Beroe.

met with in summer on most parts of the coast, swimming near the surface, and may readily be taken in a gauze drag-net. It has a melon-shaped body, from half an inch to nearly an inch in length, clear as crystal, and divided, as it were, into gores by eight longitudinal equidistant bands or ribs. These ribs, when minutely examined, are found clothed with innumerable flat plates resembling the paddles of a water-wheel placed one above another, and acting under the control of the will of the animal. When the Beroe wishes to move, these paddles are set in motion, and by their united action on the water propel the living globe of crystal, with a swift and easy motion, forward or backward, as it wills; and when it wishes to turn, it merely stops the movement of the paddles on one side. The cilia, in sunlight, reflect brilliant prismatic colors, and in darkness flash with a beautiful blue light. Delicate as are its organs of motion, the fishing apparatus of the Beroe is not less elegant. This consists of two long and exceedingly slender tentacula, five or six inches in length when fully extended, but capable of being wholly drawn within the body of the creature, where they are lodged in tubular sheaths. To the long filament is attached, at regular distances, a multitude of shorter and much more slender fibres, which are coiled up in spirals when the main filament contracts, and gradually spread out as it lengthens. These are very similar to the small hooked threads attached at intervals along a fishing-line."

Questions.—What is said of the symmetry of the Radiates? What of the structure of the Actiniæ? What is the office of the chambers around the stomach? What are the type-families among the Radiates? What is said of the locomotion of the Radiates? What of their resemblance to vegetables? What of their senses and nervous system? What are their cilia? Describe their mode of action. What purposes do they effect? What are the three classes of Radiates? What is the structure of the Starfish? In what way does it walk? How does it take its prey? What is said of the motion of its arms? What is said of the order Stellerida? What is the structure of the Echinus? How does it walk? What is the plan of its shell, and how does it grow? Where are the Echini found, and what are their habits? What is said of their masticating apparatus? What is the mode of their development? What is said of two aberrant orders of Echinoderms? What animals constitute the second class of Radiates? What are their peculiarities? What is said of the Medusæ? How are they like the Actiniæ, and how unlike them? Where are the Acalephs generally seen? What is said of the phosphorescence of the sea? What is said of the Beroe?

CHAPTER XXXV.

RADIATES — *continued.*

598. WE come now to a class of Radiates including animals which are, with some few exceptions, entirely different from those of the classes already considered in relation to locomotion. Most of the Echinoderms crawl; some of them, and all the Acalephs, swim; but the Polypes are, for the most part, like plants, fixed to the spot where they begin life. The older botanists described these animals as plants, and arranged them with seaweeds and mosses. The Sea Anemone was considered a flower, and the analogous beings found in coral and mad-

repore were spoken of as *blossoms of stony plants*. It is now about a century since their animal character was really admitted by naturalists; and it is only quite recently that their structure and habits have been thoroughly investigated.

599. The Polypes, or Zoophytes, have the most simple construction of all animals, but they differ from each other in the degree of their simplicity. The most simple of all are the Hydras—little Polypes which you can find in stagnant waters. In Fig. 270 you have a representation of one of these. The smaller figure shows it of the natural size. It is a simple sac or purse-like animal, with a mouth, and tentacula arranged around the mouth. With these tentacles or arms the animal catches its prey, and puts it into its stomach through the mouth, *a*. In its general shape, and in the working of the arms, it is much like the Cuttle-fish (§ 550). Its tentacles are, however, armed in a very different manner. They have neither suckers, like those of the Cuttle-fish, nor cilia, like those of many animals, but minute bristles, and sharp, firm spines, curiously arranged. These spines are concealed in wart-like processes when they are not in use, but they can at any time be thrust out, just as the claws of a carnivorous animal are protruded from their concealment when their services are needed.

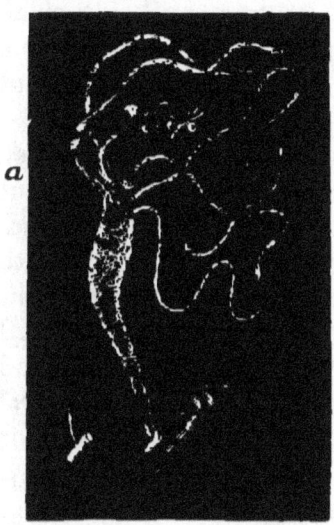

Fig. 270.—Hydra.

600. When the Hydra is searching for prey, it allows its tentacles to float about in the water, its body being fastened by a sucker to some solid substance. If a Crustacean or an aquatic worm happens to hit one of them, the arm is immediately thrown around it, as you see in

the figure, the spines being forced out to make sure the hold. If the animal caught be of sufficient size to require it, the other arms are thrown around it also, and the victim is conveyed to the stomach. It has been observed that soft-bodied animals, if held for a little while in the arms without being swallowed, always die, even when released alive; from which it is inferred that the spines convey a poisonous secretion into the bodies of the prey, as do the fang of a serpent and the sting of a bee. As the Hydra can not do this to Crustaceans or any hard-shelled animals, they do not die at once on being swallowed; and so thin is the texture of the Hydra, that the outlines of these animals can be seen as they move about inside.

601. The Hydra has some power of locomotion. When it wishes to change its place, it does it with a movement like that of the Geometrical Caterpillars (§ 479). Bending its body forward, and taking hold either by its mouth or its tentacles, it raises its sucker, and advances it. Then, fastening itself again by this, it carries forward again the upper part of its body and the tentacles, and thus slowly moves to the desired spot. It takes several hours to march two inches in this way, and seven or eight inches may be regarded as a good day's journey. But sometimes the Hydra gets along faster by executing a series of somersets, fastening himself by his tentacles, and then throwing his body forward. It sometimes, also, manages to sail along by a curious contrivance. It raises its flat sucker above the surface of the water, and letting it become dry, it acts as a sort of float, the animal hanging down in the water. In this way it can sail over considerable distances, either carried along by the wind blowing on the float, or by the tentacles acting as paddles. Though there is little of positive sensation in this animal, and therefore but a low degree of enjoyment, it undoubtedly considers this ingenious way of sailing as one of its best sports.

602. The Hydra is nothing but a stomach with tentacles attached to it. It can be turned inside out like a glove, and fare as well as before, showing that there is little, if any difference between what may be called its skin and its inside lining. Trembley, the first discoverer of the Hydræ, once witnessed a very singular circumstance: "Two Polypes had seized upon the same animal; both had partially succeeded in swallowing it; when the largest put an end to the dispute by swallowing its opponent, as well as the subject of contention. Trembley naturally regarded so tragical a termination of the affray as the end of the swallowed Polype's existence; but he was mistaken; for, after the devourer and his captive had digested the prey between them, the latter was regurgitated, safe and sound, and apparently no worse for the imprisonment."

603. Hydras are produced in two ways. One is by seeds or eggs. These are thrown out by the animal in the autumn in the form of gelatinous globules, and in the following spring Hydras come from them, and, fastening themselves to some stick or other solid substance, begin their quiet but predaceous life. Another mode of production is by buds, thus allying these animals in a marked manner to plants. Buds at first appear as slight projections from the outer surface of the body, and these gradually become perfect animals, at length separating from the parent to attach themselves to some solid body. The stomach of the young Polype communicates with that of the parent so long as they are connected together; and yet it is not uncommon to see both struggling for the same worm, and gorging opposite ends of it. Sometimes the young Hydra has buds start out from its body before it has separated from the parent, so that we have three generations in one group. This production of different generations is so rapid in some cases, that it is calculated that above a million descendants come from one animal in a month.

604. But the most remarkable fact in regard to the Hydra is, that if a small piece of its body, or even if a tentacle be torn off, the separated part will itself become a perfect animal. Thirty or forty Hydras may be produced by cutting a single one into pieces. The Hydra of ancient fable seems thus to be realized in nature. The Hydra does not seem to suffer at all from mutilation, but young Polypes sprout abundantly from any wound that may be made. Two Polypes may even be grafted together by their cut surfaces. This can be done not only with those of the same species, but with different species, as the green and brown Hydras.

605. There are some Polypes, belonging to the same order with the Hydras, which have a much stronger resemblance to plants in their habits and arrangements. They are situated on horny stalks, and, in some cases, these stalks have branches, with cells on them, for containing the little Polypes, as seen in the Sertularia, Fig. 271. The stalk and branches here are hollow, being lined with a membrane which is the essential part of the animal, or, rather, of the community of animals thus united together. Each individual Polype may be considered as having a stomach of its own, but communicating with a sort of stomach common to them all, which lines the branches and the stalk. There is in this respect an

Fig. 271.—Sertularian Polypes.

analogy to the Hydra during the temporary connection of the young Hydras with it, their stomachs having a communication with the stomach of the parent. These beautiful and delicate animals were formerly supposed to be vegetable, and were called by naturalists sea mosses.

606. Of the order of Polypes called Helianthoida I have already noticed quite particularly one group, the Actiniæ (§ 576), as illustrating well the characteristics of the Radiata. There need to be added here to what has been said some farther statements in regard to their structure and habits. It is the beauty of the expanded disks of these and other allied animals that gives the name Helianthoida to this order, this word being derived from two Greek words meaning *sun* and *form*. In the tropics they are peculiarly brilliant, and many travelers speak most enthusiastically of the gorgeous spectacles which groups of them often present.

607. Some Actiniæ live on smooth sands, spreading out their tentacles for prey, and retiring beneath the sand when danger threatens. But most of them attach themselves to rocks, often adhering so firmly that they can not be detached without lacerating them. And when portions of the disk are left fixed to the rock, new animals will be formed from them, just as is the case with sections of Hydras (§ 604). There is one species that fastens itself to some shell; and it is observed that the Hermit Crabs are fond of taking up their abode in such shells, making a singular sort of partnership.

608. The muscular structure in some of the larger species is very distinct, and exhibits great power in action. They can not only master small shellfish and Crustacea, but even Crabs, Prawns, and other Crustacea of considerable bulk. The mouth is capable of wide distention, so that animals can be taken in which one should suppose to be inadmissible. It is amusing to witness the struggles of some animal that has, in walking about, come over one of these gaping mouths, as it is caught by the tenta-

cula and thrust down into the capacious stomach. So voracious are these animals that they will attempt to swallow articles which their stomachs can not possibly accommodate. In this case the animal will perhaps hold the mass partly in and partly out of the stomach firmly with its tentacles, pushing it farther in as fast as the lower part of the mass is digested.

609. In § 599 I spoke of the arrangements of the tentacles of different animals. The structure of the tentacles of the Actiniæ is very peculiar. Their power of holding on is owing to a multitude of cells, in which there are coiled up in a spiral form fine wire-like filaments. These can be shot forth from their cells to a considerable length, and this being done with a multitude of them enables the animal to hold on fast to its prey.

610. Some of the Polypes of this order have a skeleton. It is formed inside of the animal at its lower part, and it is fastened to the spot where the Polype lives. We may consider it as a foundation frame-work for its body. Resting on this, it puts forth its arms continually to take its food.

611. But this skeleton differs from the skeletons of all other animals in one respect. Other animals retain their skeletons all their lifetime; but the Polype does not. It is constantly making new skeleton. It is a singular process, and I will describe it to you with its results. The very lowest part of the Polype is continually dying, and with it the skeleton which it covers. But as this dies the animal keeps its full size, for the body is continually supplied with new living substance on the borders of the dying portion. It grows just as fast as it dies. It therefore is all the time moving upward, making new skeleton, and leaving the old below. The result, you plainly see, would be a column of dead skeleton with the Polype at the top of it. In this column, after a while, the living part is but small in comparison with the dead part below.

612. This result you see represented in Fig. 272, one

Fig. 272.—Caryophyllia.

of the Caryophyllia. Here are two stony columns formed by two Polypes. The animals are ever at the summits, with only a small portion of the columns in their bodies and living. The rest is like dead bone. It differs from the bones of common animals in its composition. Their bones are made of *phosphate* of lime, while the Polype's skeleton is made of the *carbonate* of lime, or chalk, like the shells of the Mollusks. All of this stony substance forming these columns is supplied from the blood of the Polype. It gets into the blood from the water, and from the food which the Polype eats. The immense masses of coral seen in some localities are formed there in the same way essentially with the bones of the Vertebrates and the shells of the Mollusks and Crustacea. You observe in the figure that on the summit of one of the columns there are two Polypes, one being larger than the other. Here is the beginning of a branching process which is very common. A second Polype has started out of the side of the original one; and, as the growth and death go on, now there will be two columns instead of one from that point. And as these grow upward, there may be still other divisions in the same manner.

613. Some species of coral-forming Polypes, instead of being on branches, are distributed over a continuous surface of a stony or calcareous mass. This arrangement is represented in the Astrea Viridis, Fig. 273 (page 352). Here is a rounded mass of limestone, made up of the united skeletons of Polypes. Over its upper portion is a fleshy covering connecting the Polypes together, making what is called a polypidom, or household of Polypes. At *a a* are the Polypes, out of their cells and fully expanded. At *b b* the animals are within the cells. At *c* is

Fig. 273.—Mass of Astrea Viridis.

the stone uncovered by the flesh. Among the expanded Polypes are seen two which are out of their cells, but their tentacles are not expanded.

614. It is chiefly by the coral-forming animals of this order that the coral reefs and islands have been built. So immense are the works which large companies of these animals perform here and there, that we may regard the changes which they produce as among the most important to which the earth has been subjected, at least since it has been inhabited by man. A large number of the Polynesian Islands, and many of those in the Indian Ocean, have been constructed by these little animals. They are continually building extensive reefs, also, in various forms and in different positions. Off the coast of New Holland there is a coral reef over one thousand miles in length. Great as are the changes now going on from the agency of these little architects, it is supposed that in what may be called the forming ages of our earth they had a still greater agency, in the formation of the limestone rocks which constitute so large a part of the crust of the globe.

615. There is another order of Polypes called the Asteroida, from the star-like appearance of the tentacles. The Red Corals, the "Organ-pipe Corals," the Sea Fans, etc., belong to this order. Some in this order verge toward the sponges. Their habits are, for the most part, so much like those of the other Polypes that I will not dwell on them.

616. The proper place of the Sponges it is difficult to determine. If they are really animals, they are of the lowest grade, exhibiting not the least signs of sensation.

They consist wholly of a substance which is considered, from the smell produced by burning it, to be much like the *horny* substance found in many animals. There are two kinds of pores—a vast number of minute pores, and here and there larger ones among them, termed vents. Examined in their living state, it is manifest that from the larger pores of the Sponges water is constantly passing out in currents, and it is supposed that it as constantly passes in through the minute pores. This is analogous to some movements that occur in certain animals. The net-work of which sponge is composed is found, by examination with the microscope, to be made up of fine tubes. One hundred and fifty different species have been described by Lamarck.

617. In Fig. 274 is a representation of a section of a

Fig. 274.—Section of living Sponge.

piece of sponge, exhibiting the branches which conduct the water from the minute interstices to the large vents. The currents which come out from these vents are rendered apparent by the minute particles of matter which happen to be in them, as represented in the figure. The Sponge lives on the water and what the water holds in solution, and for its growth it is therefore necessary that water should be constantly circulating through it in the manner which I have described. There is one species in which, the Sponge being of the shape of a bottle, the absorbing pores are all on the outside, while the vents are inside. The result is that there is a strong current of water constantly pouring out of the mouth of the bottle.

Questions.—How do the Polypes differ from the other classes of Radiates in regard to locomotion? Why were they so long supposed to be plants? What is said of their construction? What is said of the structure of the Hydra? What of its mode of taking its prey? What of its locomotion? What is stated by Trembley? Describe the two ways in which Hydras are multiplied. What is said of mutilating them? What of uniting two together? What is said of the Helianthoida? Where are the Actiniæ commonly found? What is said of their multiplication from portions of their disk? What is said of their muscular structure? What of their mouths? What of their voracity? What is the structure of their tentacles? What is said of the skeletons which some Polypes have? How does their composition differ from that of the skeletons of common animals? In what other respect do they differ? How is the formation of the skeleton column exemplified in the Caryophyllia? Whence comes the supply of the material to make this skeleton? What is said of the associated Polypes as exemplified in the Astrea Viridis? What is said of the formation of the coral reefs and islands? What of the agency of the coral animals in the forming ages of the earth? What is said of the Asteroida? What is said of the structure of the Sponges? How many species are there? Describe the arrangement of the Bottle Sponge.

CHAPTER XXXVI.

CONCLUDING OBSERVATIONS.

It is my intention in this chapter to retouch some points which have been treated of, and also to bring out some others which may add to the interest of the general subject.

618. The pupil has observed, as he has proceeded, the adaptation of each animal to its circumstances and to its mode of life. This has been seen both in classes of animals and in individual cases. I will refer to a few examples of this adaptation in classes. Birds are fitted in both their internal and external structure (as you saw in

the first part of Chapter XII.) for flight in the air; while the fishes are so constructed (§ 353–357) as to swim easily in the water. And then, in those classes of birds that are designed in part for life on the water, there are special provisions for swimming in their webbed feet and other arrangements (§ 291). Some animals are carnivorous, while others are herbivorous, and others still eat a variety of food, and may even be omnivorous, like man. The adaptation of organization in these different cases has reference, as you have seen, both to the kind of food and to the mode of obtaining it. If it had reference merely to the former, it would be seen only in the teeth, the jaw, and the stomach. But in its reference to the latter, it is observed in the structure and arrangement of the organs of the senses, and even of the whole frame. For example, in the carnivorous animal of prey, there must be a full development of the senses of sight, hearing, and smell; a frame capable of quick movement; strong claws, worked by stout muscles, to hold the prey; teeth fitted to tear it in pieces, and a stomach altogether different from that of the herbivorous animals.

619. The adaptations in relation to temperature are very interesting. Animals that live in cold climates have coverings which differ greatly from those of animals living in warm countries. The elephant, with his scanty hairs, is in strong contrast in this respect with the shaggy-coated bear. Our supply of furs comes from northern regions, from animals that could not withstand the cold without such coverings. As the horse is a native of a warm climate, he requires the blanket in our winters, and for the same reason the cow and ox need to be better sheltered than is ordinarily done among the farmers of temperate climates. In the arctic regions, even animals that are protected by a furry covering have also, as a farther defense against the cold, a good layer of fat, which not only keeps the heat in by its non-conducting property, but also aids in the production of heat. We

may notice in this connection, in the insect world, the special provisions against the cold in the cocoons which are to remain through the winter to another season (§ 413).

620. The individual adaptations seen in the different species are endless in variety. Those which I have brought to your notice, in passing through the four subkingdoms of the animal world, are exceedingly few in comparison with all that might be gathered up, and new ones are coming to view every day in the researches of zoologists. Each species has its peculiar habits, and, of course, its corresponding adaptations in its structure. The study in this respect has no end, and the fertility of the wisdom and skill of the Deity is seen to have no bounds. The humblest observer who enters this field may find many things that no one has yet recorded, and thus may be a contributor to zoological science.

621. Of the individual adaptations I will notice a few of those only which are of a marked exceptional character. The whale is a Mammal having lungs, and yet it lives in the water like the fishes. For this it must have an especial adaptation in the arrangement of the circulating system, as described in § 187. So also, as it is a warm-blooded animal, its heat must be kept from escaping too rapidly by a special provision, and this must be in consonance with its fish-like habits (§ 186).—The bat is a Mammal, and yet, as it is destined to get its livelihood on the wing and in the dark, it has peculiarly constructed wings for this purpose (§ 58, 59, and 60).—Most fishes are shaped with reference to ease and rapidity of movement (§ 353). Hence they are like boats for racing, long, spindle-shaped; and they have no projections like a shoulder to prevent their gliding swiftly through the water. But there are some exceptions, as in the short, big-mouthed Lophius (Fig. 172).—Its habits explain the reason of the exception. The brain of man is but the fortieth or fiftieth part of the weight of his whole

body, and yet it receives about the fifth or sixth part of all the blood in circulation, simply because the amount of thinking done there requires this supply to keep the instrument of thought in good condition. To prevent this great amount of blood from flowing too rapidly and forcibly into the brain, the arteries, as they enter the skull, are so arranged that the flow shall be circuitous rather than direct. Then, again, there is a farther special provision against the too free admission of blood into the brain in animals that hold their heads downward much of the time, as grazing animals. When we hold our heads downward, very uneasy sensations are soon produced from the undue amount of blood in the head; but in the grazing animal this effect is prevented by a division of the arteries into a net-work before they enter the brain. In this connection I will also refer you to the remarkable provision against a sudden rush of blood to the head in the deer when the circulation in the "velvet" is stopped (§ 164).

622. The adaptations which we witness in the different conditions of animals that pass through a full metamorphosis are of exceeding interest. That the adaptations of a crawling worm should all be exchanged, during the sleep of the animal, for those of a beautiful flying insect (§ 405), is one of the most wonderful things in nature. Still more wonderful is the change of adaptation, when an animal fitted to live like a fish experiences, in the midst of a state of full activity, internal changes which prepare it at length to emerge with lungs and wings, leaving its skin behind it in the water, as exemplified in the mosquito (§ 502).

623. But adaptation is displayed in the most interesting manner in the relations of organization to the capabilities of animals. The more an animal knows, the more complicated is its structure, or, in other words, the more extensive is the machinery which is provided for the use of its mind. We see this both in the apparatus of the

senses and in that of voluntary motion, and also in the nervous system, by which these two kinds of apparatus are connected with the mind.* In the lower orders of animals the senses are very imperfectly developed, and in some most of the senses are absent. Thus, in the Hydra (§ 599) and in the Actiniæ (§ 576) there is no evidence of the existence of but one sense, that of touch. The Actiniæ are, indeed, sensibly affected by light, but this does not prove that they see. As we go upward in the scale we find the apparatus of the senses generally more and more developed. Taking all of them into view, the senses are best developed in man, though some of them, for special purposes, have a higher capacity in certain animals than in him. Some may have a more acute smell, as the dog, or see farther, as the eagle; but no animal has *all* the senses in such perfection as man. The same can be said of the muscular apparatus. The variety of muscular action is greatest of all in man, while in some animals there are special muscular endowments for special purposes above any thing of the kind to be found in him. The gradation in the nervous system is still more definitely marked. In man it has its fullest development; and, as we go down in the scale, we at length come to animals that have no distinct brain, and finally to those in which, as the hydra, no trace of any thing like a nerve can be found. In these last nervous matter is presumed to exist because actions are performed which, in animals of a more defined organization, are known to be dependent upon nervous agency.

624. Amid all the variations of structure to suit the different wants and capabilities of animals, the Creator has adopted certain general plans, so that order prevails throughout all the extreme variety of the animal kingdom. We can see this whether we take into view large

* For the relations of the senses, the muscles, and the nervous system, I refer you to the chapter on the Nervous System in my "First Book in Physiology."

groups, as classes or sub-kingdoms, or smaller ones, as families or genera. It is in the typical forms that we have these plans fully brought to view; while there is in the aberrant, in proportion to the degree in which they are so, a departure from these plans, or, rather, a modification of them, to suit the particular wants and habits in each case. Thus, in the Vertebrates, the plan of the skeleton is very perfectly developed in the higher animals, and especially in man. But the general features of the plan are the same in all this sub-kingdom. This may be seen if we take the skeleton as a whole, as illustrated in the first chapter, or if we look at some particular portion of it, as the arm and hand, as illustrated in regard to the flipper of a Whale (§ 185), the anterior extremity of the Dugong (§ 195), the wing of the Bat (§ 58), and the wing of birds (§ 198). In the Articulata, the ring-like arrangement, seen so decidedly in most of the animals of this sub-kingdom, as the Centipede (§ 381), is not really given up in those where it seems to be, as in the Crab tribe; but a careful observation shows that it is only modified by making some of the rings exceedingly broad, while others are made exceedingly narrow (§ 526). There is not here an abandonment of the general plan, but a departure or aberration from it to some extent, making an aberrant form, in distinction from the typical forms where the ring-like arrangement is fully carried out. What I have thus said of the Vertebrates and the Articulates is essentially true of all parts of the animal kingdom.

625. The great wonder is that so much uniformity of plan can be made consistent with such extreme variety, the minutiæ of exact adaptation being in all cases fully carried out. There would have been a much smaller display of wisdom and skill, if the same variety had been attained without the extended general plans which we see were adopted. None but omnipotent power could so connect endless variations in minutiæ with so few typical forms and arrangements.

626. With these general plans there is in every animal a marked relation of each part to every other part. Every bone, for example, not only has its exact relation to every other bone, but also to every other part and organ. It is from this harmony existing in every animal frame that the zoologist is able to know the general structure and habits of an animal on inspecting a single bone or tooth belonging to it. For example, suppose that he picks up a tooth with two stout roots and a sharp cutting edge rising to a point, such as you see in Fig. 275. Let

Fig. 275.

us see what he can know in regard to the animal to which this tooth belonged. First he would know that it was a Vertebrate, for no teeth at all like this are ever found in an animal outside of the Vertebrate classes. He knows, therefore, that this animal had a brain and spinal marrow, that its senses were well developed, and that its blood was red. Then the two long roots show that the tooth was deeply implanted in a double socket, and that the animal was, therefore, a Mammal, for this arrangement is seen only in that class. The cutting edge of the crown indicates that the animal was a carnivorous quadruped, and that its jaws moved upon each other with a scissors-like motion (§ 67), and not a grinding one, as in the herbivorous quadrupeds. It may be inferred, also, that the feet were not hoofed, but armed with claws for securing the prey, and that the muscles both of the limbs and head were very strong. The general shape of the animal (§ 70) can also be made out, and its size can be estimated from the size of the tooth. The kind of stomach which it had can also be known (§ 68). Baron Cuvier had great skill in such studies. From a single bone, or even a piece of one, he could picture an entire skeleton, and describe the character and the habits of the animal.

627. The general plans adopted by the Creator should,

of course, be our guide in the classification of animals, so that it may be a natural and not an artificial classification. In studying nature we should always endeavor to read correctly the traces of the mind of the Creator.

628. The distribution of animals in the various regions of the earth is a very interesting subject, but my limits will allow of but a brief notice of it. Man is the only animal that is found in every part of the earth. He is thus a cosmopolite, because he has a mind that can contrive clothing and habitations suitable to every variety of climate. Next to him in general diffusion are some of those animals which are domesticated by him, and also some which follow him and dwell in his habitations, as the mouse, the rat, the fly, etc. Most animals are limited to certain regions, differing, however, in the extent of their diffusion—some having a wide range, while others are confined to comparatively narrow limits. Those animals which are found in any particular region or country are said to constitute its Fauna, as the flowers found there make up its Flora. We speak of the Faunas of the arctic, the temperate, and the tropic regions. Then, also, we subdivide these into Faunas of portions of these regions of greater or less extent, according to circumstances. The dividing lines between the different zoological provinces thus marked out are by no means impassable boundaries, for there is generally a mingling of animals near the borders of two adjacent Faunas. Thus, although the Fauna of the United States and that of the region west of the Rocky Mountains are very distinct, yet these mountains do not effect an entire separation, for some animals of either Fauna are found on both sides of the range.

629. The Faunas of the arctic region have comparatively few species, but the number of individuals of each is often immense. Especially is this true of the fishes and the birds. The birds are mostly of the aquatic tribes—gulls, cormorants, ducks, petrels, etc. All the animals

are of a dull color. Not a bird of bright plumage is to be found. Of terrestrial animals, the most noticeable are the White Bear, the Reindeer, the White Fox, etc.; and of the aquatic Mammals, the Seals and the Whales. There are no reptiles, few insects, and no coral animals.

630. In the Faunas of the temperate regions there is much greater variety than in those of the arctic. Terrestrial animals abound here. The birds exhibit considerable variety of color. One of the prominent features of the Fauna of the temperate zone is the constant change which is going on in it from the variety in the seasons. Especially is this true of the northern portion. In the colder months insect life has retired for hibernation, and vegetable life is, for the most part, in a similar state. The sources of livelihood for many animals are thus cut off. The birds, therefore, migrate to warmer regions, and many of the mammals hibernate; and in the spring the mammals wake up, and the birds return, making nature, which was so still in winter, vocal again.

631. Abundance, variety in form, and brilliancy of colors are the distinguishing characteristics of the tropical Faunas. "All the principal types of animals," say Agassiz and Gould, "are represented, and all contain numerous genera and species. We need only to refer to the tribe of Humming-birds, which numbers not less than 300 species. It is very important to notice that here are concentrated the most perfect, as well as the oldest types of all the classes of the animal kingdom. The tropical region is the only one occupied by the Quadrumana, the herbivorous Bats, the great Pachydermata, such as the Elephant, the Hippopotamus, and the Tapir, and the whole family of Edentata. Here, also, are found the largest of the Cat tribe, the Lion and Tiger. Among the Birds, we may mention the Parrots and Toucans as essentially tropical; among the Reptiles, the largest Crocodiles and gigantic Tortoises; and, finally, among the articulated animals, an immense variety of the most beautiful insects.

The marine animals, as a whole, are equally superior to those of other regions; the seas teem with Crustaceans and numerous Cephalopods, together with an infinite variety of Gasteropods and Acephala. The Echinoderms there attain a magnitude and variety elsewhere unknown; and, lastly, the Polypes there display an activity of which the other zones present no example. Whole groups of islands are surrounded with coral reefs formed by these little animals."

632. This variety is made more striking by the fact that each continent has many animals in its tropical region peculiar to itself. Thus the Giraffe and Hippopotamus appear only in Africa; and that strange animal, the Sloth, is found only in America. The 300 species of Humming-birds are exclusively American, nearly all of them being tropical. The Sunbirds, on the other hand, which are somewhat like them, do not appear at all in America, but are widely scattered over Asia, Africa, and the islands of the Pacific.

633. Some of the local Faunas have prominent peculiarities. The Fauna of Brazil is exceedingly rich, with its gigantic Reptiles, its Monkeys, its Edentata, its brilliant Humming-birds, and its wonderful variety of insects. There is no part of the world that has so peculiar a Fauna as Australia. Here are great numbers of Marsupial animals. Here, also, is that strange animal, the Duck-billed Platypus (§ 133); and here, too, is the Black Swan, supposed to be an impossibility till it was found in that singular country.

634. The pupil has by no means obtained an adequate idea of the abundance and variety of the animal kingdom from what he has seen in this book of its different departments. In so small a space only a few specimens of each group could come under consideration. That you may have some idea of the extent of the field which zoology has opened, I will give you some statements of the numbers of animals from Agassiz and Gould. The number

of species of Vertebrates is probably 20,000, of which the Mammals are 2000, the Birds 6000, the Reptiles 2000, and the Fishes 8000 or 10,000. There are probably over 15,000 Mollusks. The Insects are the most numerous class of animals, there being already collected from 60 to 80,000 species. Of all the Articulates there are about 100,000 now known, and it is safe to compute the whole number at 200,000. If we add to the above 10,000 for the Radiates, we shall have about 250,000 species. It is also estimated by Agassiz that there is about the same number of species of fossil animals; that is, those which are not now in existence, but which are known to have existed by the remains that we find of them in the rocks and in the earth. I have noticed a few of these in passing, as the Mastodon (§ 139), the Iguanodon (§ 326), and the Ammonites (§ 551).

635. But farther than all this, we can get no adequate idea of the abundance of animal life if we do not take into view the minuter living forms, as well as those which are ordinarily noticed. These I have not considered, because it would lead me into too wide a field. Quite large portions of the earth—of its rocks, and mountains, and sand, and mud, and dust—are made up in part of the remains of *minute* animals, called, therefore, animalculæ, or, in English, animalcules. Some of these are so small that their structure can not be made out except by the aid of the microscope, and some can not even be seen at all by the naked eye. For example, the stone used for building in Paris, and in all the country round it, is so full of the shells of an animalcule, that there are 58,000 in a cubic inch, or three thousand millions in a cubic yard. This animal belongs to a group which are called Foraminifera, because their shells are full of little foramens or openings. The substance within the chambers of the shell is mostly a translucent jelly, and through the openings branch out root-like legs, on which it is curious to see the animal walk. Foraminifera, perhaps of

the size of the head of a small pin, may sometimes be seen thus walking on the glass walls of an aquarium; and a great variety of species can be found in the sand of most sea-coasts, as any one may see if he examine a handful of it with a pocket lens. In the chalk formations there are remains of even smaller animals than these. Ehrenberg, on examining chalk very minutely divided, found in it some many-chambered shells, some of them whole and some in fragments. He calculated that there were a million in every cubic inch, or ten millions in every pound. He was able to discern them even in the glazing of a visiting card, although the chalk in this case had been subjected to such minute division that one would suppose all trace of organization to have been lost.

636. The earth in and about the city of Richmond, Virginia, is filled with various shells of Animalculæ. A portion of one of these shells, as seen through a powerful microscope, is given in Fig. 276. There are various species of this shell, called, very appropriately, Coscinodiscus (sieve-like disk), varying in size from the one hundredth to the one thousandth of an inch in diameter. The guano brought from the island of Ichaboe is found to contain multitudes of this and other shells, making a beautiful display as a little of the dust is placed in the field of the microscope. These shells are the remains of animalculæ that lived in the water and were eaten by fishes. Then these fishes were devoured by sea-birds; so that these shells must have passed through the process of digestion twice, and after that were exposed in the guano-bed to the ordinary causes of decay perhaps for centuries; and yet, says Professor Brocklesby, "under all these influences they remain unchanged, and the eye of the naturalist at last detects these minute structures, still possess-

Fig. 276.

ing their original beauty, with the delicate tracery of their rich configuration, almost as sharp and clear as it was, perhaps, a thousand years ago."

637. The Tripoli, or rotten-stone of Bohemia, which, when ground, is used as a polishing powder, is full of flinty shells, which are so minute that forty thousand millions are contained in a single cubic inch. Other instances, in great number, could be cited, from various quarters of the world, of large deposits of the remains of animalcules, in rocks, in earth, in peat-bogs, and in mud. Well does Lamarck say of these deposits, that "it is by means of the smallest objects that Nature every where produces her most remarkable and astonishing phenomena. Whatever she may seem to lose in point of volume in the production of living bodies, is amply made up by the number of individuals, which she multiplies with admirable promptitude, to infinity. The remains of such minute animals have added much more to the mass of materials which compose the exterior of the crust of the globe than the bodies of Elephants, Hippopotami, and Whales." In § 614 I spoke of the agency of coral animals in building up portions of the earth by the formation of their skeletons; but the agency of these animalcules, by means of their remains, is vastly greater.

638. The name Infusoria was given to animalcules because they abound in infusions of decomposing vegetable or animal substances. By some, however, this term is confined to those animalcules which have cilia, by which they swim through water. An abundance of these can be obtained in warm weather from the surface of water in ponds, especially where there is a reddish or green tinge, or a slimy layer. In Fig. 277 you have a variety of these Infusoria. They move about very freely in the water by means of their cilia. "These movements," says Carpenter, "are extremely various in their character in different species; and when a number of dissimilar forms are assembled in one drop of water, the spectacle is en-

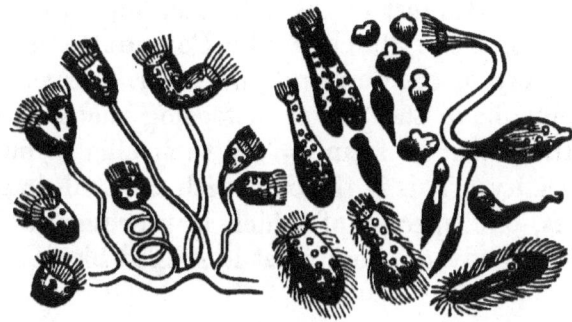

Fig. 277.

tertaining. Some propel themselves directly forward with a velocity which appears (when thus highly magnified) like that of an arrow, so that the eye can scarcely follow their movement; while others drag their bodies slowly along, like the Leech. Some make a fixed point of some portion of the body, and revolve around it with great rapidity; while others scarcely present any appearance of animal motion. Some move forward by a uniform series of gentle undulations or vibrations; while others seem to perform consecutive leaps, of no small extent compared with the length of their bodies. In some instances the body is furnished with stiff bristles and hooks, by the agency of which the animalcule is enabled to run and leap upon the stems and leaves of aquatic plants. In short, there is scarcely any kind of movement which is not practiced by these animalcules. They have evidently the power of steering clear of obstacles in their course, and of avoiding each other when swimming in close proximity. By what kind of sensibility the wonderful precision and accuracy of their movements is guided is yet very doubtful." One of the most singular of these Infusoria is the Baccillaria Paradoxa, which is composed of several parts arranged like a sliding ruler. It moves along by sliding these parts upon each other, first thrusting them forward, then closing those in the rear upon the part farthest in front.

368   NATURAL HISTORY.

639. Though most of the Infusoria move freely about in fluids, some are attached, like Polypes, to some solid base. Many of them are not, however, always thus attached, but have the power of loosing themselves from their attachment to swim off by their cilia to find some other locality. This is the case with the Bell-shaped Animalcules, one species of which is represented in Fig 278. The body of the animal is shaped like a bell, and

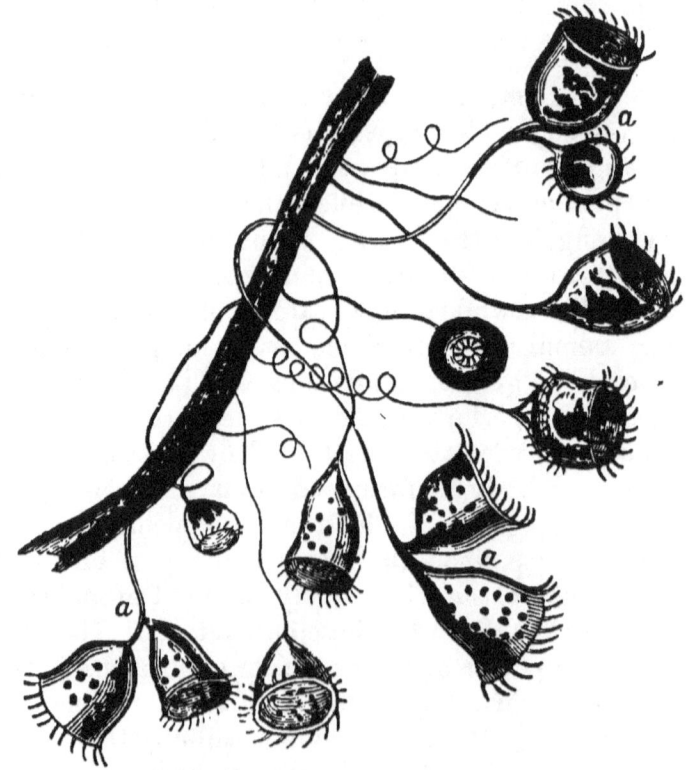

Fig. 278.—Bell-shaped Animalcules.

its margin, which is its mouth, is fringed with cilia. The actual length of its body varies, in different individuals, from the one two hundred and eightieth ($\frac{1}{280}$) of an inch to the one five hundred and seventieth ($\frac{1}{570}$). The tiny stem by which each animalcule is attached has a muscle in its whole length by which its direction and length can be altered. When the little creature is alarmed, it sinks down quickly to the place of its attachment by coiling its stem, or cable, as it may be called. In some cases, as at *a a a*, there are two animalculæ on one stem, one having grown out from the other, after the manner of some of the coral Polypes (§ 612). You see in the figure some stems without any animalcules. Here they have separated themselves from their attachment and swum away. It is an emigration to better their condition and begin a new colony.*

640. The field to which I have in this book introduced the pupil is a very broad and fruitful one, and on every side invites, in the most attractive manner, your investigation. Go, then, into the garden and the field, to the sea-side and the river-side, to the pond and the bog, and watch the movements of animals, and gather materials for observation at home. The Aquaria, now so properly becoming fashionable, furnish admirable means for carrying on some of these observations. Even with but a small portion of your time devoted to the investigation of nature, you will soon find that you do not need to go to a museum to see the wonderful and the beautiful creations of Almighty power, but that these are all around you, and even in the dust beneath your feet.

641. The animal kingdom is a great harmonious whole, with all its forms, from the minute Infusoria to the monstrous Elephants and Whales, having fixed relations to each other. These relations are not all known, but more and more of them are every day discovered. And amid all the apparent confusion and hazard attending the natural increase on the one hand, and the destruction effected on the other by the voracity of animals and other

---

* For more full information in regard to the Infusoria, I would recommend to both teacher and pupil a work by Professor Brocklesby, entitled "Views of the Microscopic World," published by Pratt, Oakley & Co.

causes, a superintending Providence maintains the general harmony, preventing any dangerous permanent increase or destruction of any species. Multitudes have, indeed, been destroyed in ages long gone by; but this was for definite purposes, which the geologist has been able, for the most part, to decipher.

642. Not only have all animals relations to each other, but they have relations, direct or indirect, near or remote, to man. The earth is his residence, and all things in it were made for him. Hence it is that in his organization the same general principles are in play which we find exhibited in the animals to which he bears the relations referred to. But while he is thus linked to the animal existences around him, he is the only animal on the earth that is destined to live any where else. He is linked to other higher existences by the possession of a soul, which has been very properly said to be "that side of our nature which is in relation with the Infinite;" and, by virtue of this, when his relation to the animals of this world ceases, another and a more glorious body is provided for him with adaptations fitted to his new and eternal condition.

Questions.—What is said of adaptation? What of it in birds and fishes? What of it in relation to carnivorous animals? How are the coverings of animals adapted to climate? What is said of the fat of animals in the arctic regions? What example of adaptation to temperature is given from the insect world? What is said of the variety of individual adaptations? How is the organization of the Whale adapted to its mode of life? What is said of the Bat? What of fishes? What is stated in regard to the brain of man and of grazing animals? What in regard to the "velvet" of the Deer? What is said of adaptation in relation to the metamorphosis of animals? What is said of adaptation in regard to the capabilities of animals? What are the relations of the senses, the muscles, and the nervous system to each other? What is said of the senses of animals? What of their muscles? What of the nervous system? What is said of the general plans of the Creator? Illustrate by reference to the Vertebrates and to the Articulates. What is there especially wonderful in the carrying out of these plans? What is said of the mutual relations of the parts of an animal? Give in full the illustration in regard to a tooth. What is said of Cuvier? How should the classification of animals be made out? Why is man a cosmopolite? What animals come next to him in extent of diffusion? What is said of the distribution of most animals? What is a Fauna? What is said of the Faunas of the arctic regions? Of the temperate? Of the tropic? What circumstance increases the variety in the tropical Faunas? What is said of some of the local Faunas? Give the statement of the numbers of species in different departments of the animal kingdom? What are fossil animals? What is said of their number? Mention some that have been noticed in this book, and give some facts in regard to them. What is said of minute animals? Give the statement in regard to the shells of Foraminifera. Describe the structure and habits of these animals. What is said of the chalk formations? What is said of the shells of the Coscinodiscus? Of the Tripoli? What of animalcular deposits? Why are animalcules termed Infusoria? To what animalcules do some confine this term? Where can these be obtained? What is said of their forms and motions? What is said of the Baccillaria Paradoxa? How are some Infusoria like Polypes? Give the statement in regard to the Bell-shaped Animalcules. What is said of the field opened to you in the observation of nature? What of the mutual relations of the animal world? What of the preservation of them by Providence? What of the relations of man to animals? What of his higher relations?

GLOSSARY.

(The numbers refer to the paragraphs where the terms may be found explained.)

Term	¶	Term	¶
Aberrant	70	Fauna	628
Acalephs	583	Felidæ	64
Acephalous	564	Fissirostres	236
Amphibious	100	Foraminifera	635
Animalcules	635	Fossil	634
Aphaniptera	419	Gallinaceous	275
Aptera	419	Ganglion	15
Aquatic	100	Gasteropoda	550
Arboreal	52	Genus	21
Asteroida	615	Haustellate	393
Aurelia	486	Helianthoida	606
Bimana	24	Hemiptera	419
Branchia	535	Herbivorous	67
Buccinidæ	562	Hymenoptera	419
Byssus	566	Imago	403
Canidæ	64	Incubation	205
Carapace	10	Infusoria	628
Carnivora	64	Insectivora	64
Cephalopoda	550	Larva	403
Cephalo-thorax	299	Lepidoptera	419
Cephalous	550	Mammal	23
Cheiroptera	24	Mandible	214
Chrysalis	486	Mantle	566
Cilia	582	Marsupial	64
Cirrhipoda	532	Mollusk	16
Class	21	Mustelidæ	64
Coleoptera	419	Nacre	554
Conchifera	564	Neuroptera	419
Conirostres	236	Nictitating	207
Crinoidea	593	Nocturnal	216
Crustacea	521	Omnivorous	93
Decapoda	532	Order	21
Dentirostres	236	Orthoptera	419
Digitigrade	92	Oviparous	23
Diptera	419	Pachydermata	64
Diurnal	216	Palpi	394
Dorsi-branchiata	535	Pedimana	24
Echino-dermata	583	Phocidæ	64
Edentata	64	Phytozoa	583
Elytra	398	Plantigrade	92

Plastron	314	Species	21
Pro-legs	479	Sub-kingdom	21
Pteropoda	550	Tenuirostres	236
Pulmonifera	560	Terrestrial	100
Pupa	402	Tunicata	564
Quadrumana	42	Typical	70
Radiate	17	Unguiculata	24
Rhynchota	491	Ungulata	24
Rodentia	64	Ursidæ	64
Ruminantia	64	Vertebra	3
Siphuncle	554		

INDEX.

(The numbers refer to the pages.)

Acalephs	341	Astrea Viridis	351
Acephalous Mollusks	328	Auks	182
Actiniæ	334, 349	Avocets	178
Aculeata	274	Axis Deer	98
Adaptation	354		
Adjutant	177	Baboons	34
Agile Gibbon	33	Babyroussa	82
Air-bladder of Fishes	211	Badgers	61
Air-cells of Birds	118	Bald Eagle	129
Albatross	184	Banxrings	68
Alligators	195	Barnacles	311
Ambergris	111	Barn Owl	137
Ambulatoria	255	Batrachia	204
American Race	28	Bats	39
Ammonites	320	Baxillaria Paradoxa	367
Amphibia	204	Bearded Vulture	136
Anchovy	221	Bears	58
Anemones, Sea	334	Beavers	71
Angora Goat	95	Bedbugs	293
Animalcules	364	Bees	278
Ant-eaters	73	Beetles	246
Antelopes	99	Bell-shaped Animalcules	368
Antennæ	230	Berenice	286
Ant-lion	267	Beroe	342
Ants	276	Birds	115
Ants, White	265	Bison	91
Aphaniptera	298	Bivalves	317
Aphidæ	291	Blackbirds	152
Apoda	221	Blind Worm	200
Apples of Sodom	271	Bloodhound	51
Aptera	298	Blubber of the Whale	109
Apteryx	171	Bluebird	148
Arachnida	299	Blue Jay	143
Argonauts	321	Blue Stocking	178
Argus Pheasant	169	Boas	204
Armadilloes	74	Bobolink	142
Articulates	19, 225	Book Lice	266
Articulates, circulation in	226	Book Scorpion	306
Articulates, nervous system of	19	Bovidæ	90
Ass	85	Bower-bird	144

INDEX.

Brahmin Bull 91
Brown Thrasher 152
Brush Turkey..................... 170
Buccinidæ 327
Buffaloes 91
Bustards........................... 173
Butcher-birds 141, 448
Butterflies......................... 285
Buzzards.......................... 183
Byssus 330

Cabinet Beetle 249
Cachelot........................... 110
Caddice Flies 268
Camelopards 107
Camels............................. 103
Canidæ 50
Canker Worms................... 284
Capabilities of Animals, adaptation in relation to.......... 357
Capricorn Beetles 250
Capridæ 95
Capybara 72
Carapace........................... 18
Carnivora.......................... 43
Carrier Pigeons.................. 168
Carrion Beetles.................. 248
Caryophyllia 351
Cashmere Goats................. 95
Cassowary 171
Caterpillar-hunters 247
Caterpillars...................... 282
Cats 49
Caucasian Race.................. 27
Cecropia Moth................... 241
Cedar-bird 153
Centipede 225
Cervidæ 96
Cetacea 108
Chalk formations 365
Chameleon....................... 196
Cheese-hoppers................. 293
Cheiroptera 23, 29
Chevaliers........................ 286
Chickadee 148
Chimpanzee 31
Chipping-bird 142
Chrysalids........................ 236
Chrysididæ 272
Cicadæ 290
Cilia................................ 337

Circulation in Brain............ 356
Circulation of Articulates..... 226
Circulation of Crocodiles 194
Circulation of Fishes.......... 209
Circulation of Mammals...... 189
Circulation of Mollusks....... 319
Circulation of Reptiles 189
Cirrhipoda........................ 311
Civet Cats 49
Clams............................. 332
Climbing Birds 161
Clio Borealis 322
Clothes Moths 288
Coaita Spider Monkey 36
Cobra di Capello 203
Cochineal 292
Cockles........................... 331
Cockroaches..................... 253
Cocoons 238
Cold-blooded Vertebrates..... 187
Coleoptera 245
Colubrine Snakes 202
Condor............................ 135
Cone-billed Perchers 142
Cone family of Mollusks...... 327
Coral.............................. 350
Cormorant....................... 185
Coscinodiscus 365
Cowries 327
Crabs 308
Cranes 175
Craw of Pigeon 166
Creepers 160
Crickets 256
Crinoidea 340
Crocodiles....................... 193
Crop of Birds 120
Cross-bills....................... 146
Crotalidæ 202
Crows 143
Crusader Carrion Beetle...... 249
Crustacea 367
Cuckoos 164
Curculios........................ 251
Cursores......................... 170
Cursoria......................... 253
Cuttlefish 320

Day-flies........................ 264
Decapoda 309
Deer 96

INDEX. 377

Dentirostres	147	Feet of Moles	67
Dipper	151	Feet of Monkeys	30
Diptera	293	Feet of Opossums	76
Divers	182	Feet of Otters	57
Dogs	50	Feet of Ruminants	88
Dolphins	112	Feet of Seals	62
Domestication	28, 51	Feet of Spiders	302
Dorsi-branchiata	312	Feet of Starfishes	338
Doves	165	Feet of Swimming Birds	179
		Felidæ	45
Eagles	127	Finches	142
Earthworm	313	Fire-flies	250
Earwigs	254	Fire Hangbird	143
Echidna	79	Fishes	208
Echinus	339	Fishes, abundance of	214
Edentata	73	Fishes, circulation in	209
Eels	221	Fishes, eggs of	215
Egg, formation of Bird in	121, 181	Fishes, shape of	210, 356
Eggs of Crustacea	308	Fishes, skeletons of	212
Eggs of Fishes	215	Fishing Hawk	129
Eggs of Insects	233	Fissirostres	154
Eggs of Shark	216	Flamingo	181
Eider Duck	121	Flatfish	220
Electrical Eel	222	Fleas	298
Elephant	80	Flies	293
Elephant Seal	64	Flipper of Whale	109
Elks	97	Flounders	221
Elytra	232	Fly-catchers	152
Emu	171	Flying-fish	211
Entellus	33	Flying Dragon	199
Entozoa	315	Flying Lemur	39
Ephemeridæ	264	Flying, mechanism of	117
Ermine	56	Foot of Mollusks	318
Ethiopian Race	28	Foraminifera	364
Eyes of Camel	104	Formicidæ	276
Eyes of Felidæ and Ruminants	90	Fossils	364
Eyes of Fishes	212	Fowls	168
Eyes of Insects	230	Franklin on American Eagle	130
Eyes of Spiders	305	Frigate Pelican	187
		Frog-hoppers	291
Falcons	126	Frogs	206
Father-long-legs	306		
Fauna	361	Gall-flies	270
Feathers, structure of	116	Gasteropods	324
Feet of Beavers	71	Gazelles	101
Feet of Camels	104	Gecko	198
Feet of Elephants	80	Geese	180
Feet of Felidæ	45	Gibbon, Agile	33
Feet of Goatsucker	155	Gilded Dandy	252
Feet of Insects	232	Gills	209
Feet of Jacanas	179	Giraffe	107

Gizzard of Birds	119	Ianthina	333
Gizzard of Insects	231	Ibex	95
Glutton	61	Ibis	177
Gnu	103	Ichneumon family of Insects	272
Goats	95	Ichneumons	49
Goatsucker	142, 154	Iguanas	198
Golden Eagle	127	Iguanodon	199
Goldfinch	143	Imago	235
Goshawk	131	Infusoria	366
Gossamer Spiders	303	Insectivora	66
Grallatores	173	Insects	225
Grasshoppers	257	Insects, digestive organs of	231
Gravedigger Beetle	249	Insects, distribution of	234
Greatfoots	170	Insects, metamorphosis of	235
Grebes	182	Insects, respiration of	228
Grizzly Bear	59	Itch animal	306
Grosbeaks	143	Ivory	81
Grouse	169		
Gulls	183	Jacanas	179
Gyrfalcon	126	Jackals	54
		Jaguar	48
Hairworms	315	Jays	143
Half-winged Insects	289	Jelly-fishes	341
Halibut	221	Jerboas	71
Hand, capabilities of	24, 29	John Dory	218
Hares	72	Jumping Insects	256
Hawks	131		
Hawksbill Turtle	193	Kalong Bat	42
Hedgehog	68	Kangaroo	76
Helianthoida	349	Katydids	257
Hemerobiidæ	267	Kingbird	152
Hemiptera	289	Kingfisher	157
Hermit Crabs	310	Kinkajou	62
Herons	176	Kites	131
Herrings	221	Kudu	102
Hippopotamus	84		
Hive Bees	279	Labyrinthic Spider	304
Honey-dew	292	Lac	292
Honey-suckers	160	Lady-birds	246
Hoopoe	160	Læmodipoda	311
Horn-bills	146	Lampreys	222
Horn Bug	249	Language of Man and Animals	27
Horned Horse	103		
Horse	84	Lapwing	174
Howling Monkeys	37	Larva	235
Humble Bees	278	Leaf-eaters	252
Humming-birds	158	Leeches	315
Hyænas	55	Legs of Birds	122
Hydra	345	Legs of Crustacea	307
Hymenoptera	269	Legs of Wading Birds	173
		Lemuridæ	37

INDEX.

Leopards	48
Lepidoptera	281
Lightning Spring Beetle	250
Limnæa Spiralis	324
Limpets	325
Lions	46
Lizards	196
Llamas	106
Locusts	258
Long-eared Bat	40
Lophius	219
Loris	38
Louse	298
Lynxes	48
Malay Race	28
Mammals	22
Man, hand of	24
Man, relation of to Animal Kingdom	24
Man, skeleton of	14
Man, superiority of, to Animals	25–27
Mandrill	35
Manidæ	73
Mantis Religiosa	254
Manyplies	89
Marmosets	37
Marsupials	76
Martins	156
Mason Spider	304
Mastodon	81
Medusa	341
Melliferous Aculeata	278
Membrane-winged Insects	269
Mermaids	114
Metamorphosis of Insects	235
Migration of Birds	123
Migration of Fishes	215
Mocking-birds	150
Mole-hills	67
Moles	66
Mollusks	316
Mongolian Race	28
Monitors	199
Monkeys	33
Moschidæ	98
Mother Carey's Chickens	183
Moths	287
Mound Birds	121
Mouse	70
Mud-wasp	275
Musk Deer	99
Musk Ox	94
Musquitoes	295
Musquitoes, eggs of	296
Musquitoes, proboscis of	296
Mustelidæ	55
Nacre	322
Naked-eyed Lizards	201
Narwhal	113
Natatores	179
Nautilus	321
Nest of Social Wasp	276
Net-winged Insects	261
Neuroptera	261
Newts	207
Night-hawk	155
Nightingale	148
Ocean Snail	333
Opossums	25, 76
Orang-outang	32
Oriole	143
Orthoptera	253
Oryx	102
Osprey	129
Ostriches	170
Otters	57
Ounce	48
Ovidæ	95
Owls	136
Ox	90
Oyster-catcher	174
Oysters	330
Pachydermata	80
Palm Weevils	252
Palpi	230
Pangolins	73
Paradise, Birds of	145
Parrots	162
Pearl Oysters	331
Pectens	331
Pedimana	29
Pelicans	185
Pellet Beetles	248
Penguins	183
Perchers	139
Perching of Birds	122
Peregrine Falcon	127

380 INDEX.

Petrels	183	Respiration of Whales	109
Pheasants	168	Rhinoceros	83
Phocidæ	62	Rhinoceros Birds	84
Phosphorescence of Sea	342	Ricebirds	143
Phytozoa	337	Robins	150
Pigeons	165	Rodentia	68
Pine Marten	56	Roe of Fishes	215
Plans in the Animal Creation	358	Rooks	144
Plant-lice	191	Rorqual	112
Plastron	191	Rotifera	315
Platypus	78	Rotten Stone	366
Plovers	174	Ruby-tailed Flies	272
Plumage of Birds	123	Ruminantia	87
Polar Bear	59	Ruminantia, eyes of	90
Polypes	344	Ruminantia, stomach of	88
Polyzoa	333	Running Birds	170
Porcupine	71	Rusty Vapor Moth	288
Porpoise	112		
Prawns	310	Sable	56
Proboscis of Elephant	80	Salamanders	207
Proboscis of Insects	229	Saltatoria	257
Proboscis of Monkey	34	Sardines	221
Pteropoda	322	Sawflies	274
Puffins	182	Scale Insects	292
Puma	48	Scale-winged Insects	281
Pupa	235	Scallops	331
		Scansores	161
Quadrupeds	43	Scavenger Beetles	248
Quails	169	Scorpions	299
		Scratching Birds	164
Raccoon	60	Sea-cows	114
Radiates	334	Sea Cucumbers	341
Rails	179	Sea Eggs	339
Raptores	125	Sea Fans	352
Raptoria	254	Seahorse	218
Rasores	164	Seals	62
Rats	70	Sea, phosphorescence of	342
Rattlesnakes	202	Sea Unicorns	113
Ravens	144	Secretary Bird	130
Rays	223	Senses of Birds	121
Razor-shell	332	Senses of Fishes	212
Redbird	143	Senses of Insects	230
Reindeer	97	Senses of Radiates	336
Reptiles	187	Senses of Serpents	201
Reptiles, brain of	190	Serpents	201
Reptiles, circulation in	189	Serpula	312
Respiration of Actiniæ	336	Sertularia	348
Respiration of Arachnida	300	Seventeen-year Locusts	291
Respiration of Fishes	208	Shad	221
Respiration of Insects	228	Sheath-bills	170
Respiration of Mollusks	319, 330	Sheath-winged Insects	245

INDEX.

Sheep	95	
Shells	317	
Shield-louse	292	
Shrew Mouse	67	
Shrikes	148	
Shrimps	310	
Sickle-bill	178	
Silkworms	238	
Simiadæ	31	
Sirens	207	
Skeleton of Bat	39	
Skeleton of Camel	15	
Skeleton of Dugong	114	
Skeleton of Man	14	
Skeleton of Ostrich	16	
Skeleton of Perch	16	
Skeleton of Turtle	17	
Skunks	56	
Sloth	75	
Slugs	325	
Snails	325	
Snake Lizards	200	
Snapping Bugs	250	
Snapping Turtles	192	
Snipes	177	
Snow-bird	142	
Snowy Owl	138	
Spanish Fly	251	
Span Worms	284	
Sparrows	142	
Spawn of Fishes	215	
Species and varieties	52	
Species, defined	21	
Species, number of	363	
Spermaceti Whale	110	
Spiders	300	
Spinal Marrow	18	
Sponge	352	
Spoon-bills	176	
Spring Beetles	250	
Springbok	100	
Squirrels	70	
Stag Beetles	249	
Stang Fishes	341	
Starfish	20, 337	
Starlings	143	
Storks	177	
Sturgeons	219	
Suctoria	315	
Sunbirds	159	
Surinam Toad	206	
Swallows	155	
Swans	181	
Swift	155	
Swimming Birds	179	
Swordfish	217	
Tadpoles	204	
Tail of Birds	123	
Tail of Crocodiles	194	
Tail of Fishes	211	
Tapir	82	
Teeth of Carnivora	44	
Teeth of Echini	340	
Teeth of Fishes	213	
Teeth of Gasteropods	325	
Teeth of Rodents	69	
Temperature, adaptation in relation to	350	
Tenuirostres	158	
Terebella	313	
Teredo	332	
Termites	265	
Terns	184	
Terricola	313	
Thorny Woodcock	327	
Thrushes	150	
Tiger Beetles	247	
Tigers	47	
Tinamous Family	170	
Toads	206	
Todies	156	
Tongue of Chameleon	197	
Tongue of Felidæ	45	
Tongue of Frogs and Toads	205	
Tongue of Humming-birds	159	
Tongue of Limpets	325	
Tooth-billed Birds	147	
Torpedo	223	
Tortoises	191	
Toucans	162	
Tree-hoppers	291	
Tridacna	331	
Tripoli	366	
Trogons	157	
Troilus	286	
Trumpeter	176	
Tubicola	312	
Tumble Bugs	248	
Tunicated Mollusks	328, 333	
Turbinidæ	326	
Turbot	220	

Turkey Buzzards	135
Turkeys	168
Tussch Silk	243
Tyrian Purple	327
Unicorns	83
Univalves	317
Ursidæ	58
Vampire Bat	41
Velvet of Deer	96
Velvet-spotted Beetle	250
Venus' Comb	327
Vertebræ	13
Vertebrates	18
Vespidæ	275
Viperine Snakes	202
Vultures	134
Wading Birds	173
Walking Leaves	255
Walking Sticks	255
Walrus	65
Warblers	148
Warm-blooded Vertebrates	22
Wasps	275
Water Ousel	151
Water Shrew	68
Wax-wings	153
Weasels	55
Web-feet	179
Weevils	251
Whale Louse	311
Whales	108
Wheel Animalcules	315
Whelks	327
Whippoorwills	155
White Ants	265
Whorl family of Mollusks	326
Wing-footed Mollusks	322
Wing of Bat	40
Wing of Birds	116
Wing of Insects	232
Wing of Wax-wing	154
Wolf	52
Wolverine	61
Woodcocks	177
Woodpeckers	163
Wrigglers	294
Yak	93
Yellow-bird	142
Zebras	85
Zebus	91

THE END.

www.ingramcontent.com/pod-product-compliance
Lightning Source LLC
Chambersburg PA
CBHW030346230426
43664CB00007BB/548